高等教育双语教学规划教材

Chemical Process Simulation
Green, Energy Saving and Precise Control

化工过程模拟
——绿色、节能与精密控制

王英龙　崔培哲　田文德　编著

化学工业出版社

·北京·

《Chemical Process Simulation》将研究生学术思维训练与过程模拟实践相结合，旨在提高研究生的科学认知与工程实践能力。本书利用 GROMACS, Materials Studio, Aspen Plus, MATLAB 等软件，从分子动力学、相平衡、稳态模拟及动态控制等方面，重点阐述化工过程模拟的绿色、节能与精密控制技术。本书共 11 章内容，第 1 章主要介绍汽液平衡和液液平衡实验数据的回归，第 2 章主要介绍离子液体相行为及其热力学性质的预测，离子液体在分离混合物方面的应用，第 3～5 章主要介绍过程强化与集成方面的实例，主要包括膜分离、热集成、热耦合、热泵隔壁塔精馏技术，第 6～11 章主要介绍了萃取精馏、变压精馏、间歇精馏及反应精馏等精馏过程的动态控制案例。

《Chemical Process Simulation》可作为高等院校化工等相关专业研究生的教学参考书，也可供从事化工过程开发与设计的工程技术人员参考。

图书在版编目（CIP）数据

化工过程模拟：绿色、节能与精密控制＝Chemical Process Simulation：Green, Energy Saving and Precise Control：英文 / 王英龙，崔培哲，田文德编著．—北京：化学工业出版社，2019.11

高等教育双语教学规划教材

ISBN 978-7-122-35724-3

Ⅰ.①化… Ⅱ.①王… ②崔… ③田… Ⅲ.①化工过程-过程模拟-高等学校-教材-英文 Ⅳ.①TQ018

中国版本图书馆 CIP 数据核字（2019）第 252643 号

责任编辑：任睿婷　徐雅妮　　　　　　　责任校对：宋　玮
装帧设计：关　飞

出版发行：化学工业出版社（北京市东城区青年湖南街 13 号　邮政编码 100011）
印　　装：三河市延风印装有限公司
787mm×1092mm　1/16　印张 14¾　字数 355 千字　2019 年 11 月北京第 1 版第 1 次印刷

购书咨询：010-64518888　　　　　　　　售后服务：010-64518899
网　　址：http://www.cip.com.cn
凡购买本书，如有缺损质量问题，本社销售中心负责调换。

定　　价：59.00 元　　　　　　　　　　　　　　版权所有　违者必究

PREFACE

Chemical process simulation refers to the chemical process data as the input value of the simulation calculation, using the process simulation software to simulate the actual production process, so as to obtain the parameters of the entire chemical process or unit operation process. The chemical process simulation does not involve any actual equipment, pipelines, and energy consumption. Only through the computer to the relevant basic equations of thermodynamics and the basic equations of the chemical unit process, the results which are close to the actual working conditions can be obtained. It is of great help to the analysis, design and transformation for the actual chemical process. It has been widely used in actual production. Aspen Plus is a very powerful process simulator for tools that model chemical processes, including chemical plants, pharmaceutical plants and refineries, which provides a relatively reliable reference for the simulation and optimization of industrial processes.

The content of this book is based on postgraduate academic thinking and engineering technology cases. Through the typical operation steps of the case and QR code video, readers can exercise the application skills of Aspen Plus software. This topic focuses on the combination of principle and practical application. It is the expansion and deepening of chemical thermodynamics and chemical engineering principles. It can be used as a teaching reference book for postgraduates majoring in chemical engineering and other related majors. It can also be used for engineering development and design of chemical processes. In view of the lack of ionic liquid related simulation and application in the current Aspen Plus books, this topic adds an explanation of ionic liquid simulation and application based on Aspen Plus, and supplys the latest distillation energy-saving separation technology and dynamic control case.

This book is based on the distillation operation in the chemical industry, combining with the actual simulation calculation case to give a detailed description, focusing on the application of green, energy-saving separation technology based on Aspen Plus. This book has 11 chapters in total. The first two chapters mainly introduce the phase behavior and thermodynamic properties of ionic liquid and the application of ionic liquid in the separation of mixtures. Chapter 3~5 mainly introduce the examples of energy-saving distillation technologies, including membrane separation, heat integration, thermal coupling and heat pump partition tower distillation technology. Chapter 6~11 mainly introduce the dynamic control cases of various distillation. The content of this book is comprehensive but easy to understand. Combined with actual production, this book is illustrated and the cases are operable.

Readers can send an email to yinglongw@126.com to get the sample questions and problem source files. By studying this book, readers can improve their understanding of cutting-edge separation technology and this book can provide guidance for practical engineering problems in chemical, petrochemical, oil refining, oil and gas fields, natural gas, fine chemicals and other related professions.

Due to the limited level of writers, this book may contain some errors, and readers are urged to criticize and correct.

CONTENTS

Chapter 1
Simulation of Vapor-liquid and Liquid-liquid Equilibrium for Binary/Ternary Systems

1.1 Introduction / 1
1.2 Data Regression of Binary System / 1
1.3 Data Regression of Ternary System by NRTL / 8
1.4 Data Regression of Ternary System by UNIQUAC / 11
References / 13

Chapter 2
Application of Green Solvents in Absorption and Extraction

2.1 Introduction / 14
2.2 Molecular Dynamics Simulation / 14
    2.2.1 Generating GROMACS Supported Files / 15
    2.2.2 Defining the Unit Box and Filling Solvent / 20
    2.2.3 Energy Minimization / 22
    2.2.4 NVT Balance / 24
    2.2.5 NPT Balance / 26
    2.2.6 Finishing MD / 27
    2.2.7 Analysis / 28
2.3 Simulation of Extractive Distillation Using the Ionic Liquid / 30
    2.3.1 Analysis of Correlation Model / 30
    2.3.2 Definition of the Ionic Liquid in Aspen Plus / 32
2.4 Simulation of CO_2 Absorption Using the Ionic Liquid / 37

2.4.1　Calculation of σ-profile Value / 38
   2.4.2　Definition of the Ionic Liquid in Aspen Plus / 43
   2.4.3　Simulation of CO_2 Capture Using the Ionic Liquid / 44
2.5　Simulation of Extractive Distillation Using Deep Eutectic Solvents / 49
   2.5.1　Definition of Deep Eutectic Solvents in Aspen Plus / 50
   2.5.2　Process Simulation / 52
References / 54

Chapter 3
Membrane Separation Process

3.1　Introduction / 56
3.2　Principle of Membrane Separation / 56
3.3　Separation of DMSO-water Using Membrane / 57
References / 64

Chapter 4
Heat-integration and Thermally Coupled Distillation

4.1　Introduction / 65
4.2　Steady-state Simulation of THF-methanol System with Heat-integration / 66
   4.2.1　Simulation without Heat-integration / 66
   4.2.2　Simulation with Partial Heat-integration / 70
   4.2.3　Simulation with Full Heat-integration / 73
4.3　Thermally Coupled Distillation Process / 76
4.4　Energy-saving Thermally Coupled Ternary Extractive Distillation Process / 78
References / 86

Chapter 5
Heat Pump Distillation for Close-boiling Mixture

5.1　Introduction / 88
5.2　Main Forms of Heat Pump Distillation / 88
5.3　Heat Pump Distillation Process of Binary System Close-boiling Mixture / 90

References / 99

Chapter 6
Energy-saving Side-stream Extractive Distillation Process

6.1 Introduction / 100
6.2 Steady-state Design of Side-stream Extractive Distillation / 100
6.3 Dynamic Control of Side-stream Extractive Distillation / 101
  6.3.1 Control Structure with Side-stream Composition/Temperature Cascade Connection / 105
  6.3.2 Control Structure with S/F and Composition Controller Connection / 105
  6.3.3 Improved Dynamic Control Structure / 107
References / 112

Chapter 7
Pressure-swing Distillation for Minimum-boiling Azeotropes

7.1 Introduction / 113
7.2 Converting from Steady-state to Dynamic Simulation / 113
7.3 Control Structures of the Process without Heat-integration / 116
  7.3.1 Basic Control Structure / 116
  7.3.2 Q_R/F_1 Control Structure / 127
  7.3.3 Control Structures of the Process with FullHeat-integration / 128
References / 130

Chapter 8
Ternary Extractive Distillation System Using Mixed Entrainer

8.1 Introduction / 132
8.2 Converting from Steady-state to Dynamic Simulation / 132
8.3 Dynamic Control of Ternary Extractive Distillation Process Using Single Solvent / 135
  8.3.1 Basic Control Structure / 135
  8.3.2 Dual Temperature Control Structure / 140
  8.3.3 Composition with Q_R/F Control Structure / 142
8.4 Dynamic Control of Ternary Extractive Distillation Process Using Mixed Entrainer / 145

8.4.1 Basic Control Structure / 145

8.4.2 Composition with Q_R/F Control Structure / 146

8.5 Comparisons of the Dynamic Performances of Two Processes / 148

References / 152

Chapter 9
Hybrid Process Including Extraction and Distillation

9.1 Introduction / 153

9.2 Solvent Selection / 153

9.3 Simulation of the Extraction Combined with Distillation Process / 155

9.3.1 Extraction Combined with Heterogeneous Azeotropic Distillation Process (LEHAD) / 155

9.3.2 Extraction Combined with Extractive Distillation Process (LEED) / 160

9.4 Dynamic Simulation of Hybrid Extraction-distillation / 164

9.4.1 Selection of Temperature-sensitive Trays / 164

9.4.2 Dynamic Control of the LEHAD Process / 167

9.4.3 Dynamic Control of the LEED Process / 174

9.5 Energy-saving Hybrid Process with Mixed Solvent / 181

9.6 Dynamics of Hybrid Process with Mixed Solvent / 185

9.6.1 Selection of Temperature-sensitive Trays / 185

9.6.2 Control Structure with Fixed Reflux Ratio / 187

References / 190

Chapter 10
Batch Distillation Integrated with Quasi-continuous Process

10.1 Introduction / 191

10.2 Feasibility of Pressure-swing Batch Distillation Based on the Ternary Residue Curve Maps / 191

10.3 Double Column Batch Stripper Process / 193

10.3.1 Design of Double Column Batch Stripper Process / 193

10.3.2 Control of Double Column Batch Stripper Process / 196

10.4 Triple Column Process / 201

10.4.1 Design of Triple Column Process / 201

10.4.2 Control of Triple Column Process / 202

References / 206

Chapter 11
Simulation of Chemical Reaction Process Based on Reaction Kinetics

11.1 Introduction / 207
11.2 Continuously Stirred Tank Reactor / 208
11.3 Simulation of Cyclohexanone Ammoximation Process / 209
    11.3.1 Steady-state Simulation of Cyclohexanone Ammoximation Process / 209
    11.3.2 Dynamic Simulation of Cyclohexanone Ammoximation Process / 209
References / 225

Chapter 1

Simulation of Vapor-liquid and Liquid-liquid Equilibrium for Binary/Ternary Systems

1.1 Introduction

Thermodynamic parameters are the foundation of simulation, and the accuracy of thermodynamic parameters decides the reliability of simulation. Aspen Plus provides many parameters that can be found in the database, such as APV100 PURE36, APV100 AQUEOUS, APV100 SOLIDS, APV100 INORGANIC. Moreover, the parameters of the physical property model are determined by the experimental data, which is used for the calculation of Aspen physical property data regression system. Physical property data regression system can match physical property model parameters with experimental data of pure component or multi-component system. Aspen Plus can regress the experimental data of any physical properties input and get the thermodynamic parameters. At present, a large number of systems lack binary interaction parameters. These missing parameters can be obtained by regression and simulation of vapor-liquid or liquid-liquid equilibrium experimental data. The objective of this chapter is to provide some methods of data regression processing, and then obtain the corresponding thermodynamic parameters.

1.2 Data Regression of Binary System

The vapor-liquid phase equilibrium experiment for the binary system of methyl isobutyl ketone (MIBK)-dimethyl acetamide (DMAC) is carried out at 101.3 kPa. The correlation regression of the experimental data is conducted by Aspen Plus to obtain the binary interaction parameters of the system. The

二元汽液相平衡
实验数据的回归

isobaric vapor-liquid equilibrium experimental data are shown in Table 1.1. The NRTL method is selected for the physical property.

Table 1.1　The isobaric vapor-liquid equilibrium experimental data of MIBK-DMAC

T/K	x_1	y_1	T/K	x_1	y_1
389.15	1.0000	1.0000	417.97	0.2510	0.5810
391.30	0.9372	0.9837	419.43	0.2299	0.5527
392.14	0.9076	0.9753	424.50	0.1599	0.4391
395.85	0.7846	0.9360	429.95	0.0940	0.2948
404.60	0.5204	0.8131	431.86	0.0740	0.2418
410.40	0.3902	0.7229	433.94	0.0520	0.1779
412.00	0.3581	0.6956	434.85	0.0420	0.1469
415.10	0.2998	0.6385	436.45	0.0280	0.1011
416.93	0.2684	0.6027	439.15	0.0000	0.0000

The simulation steps of this example are as follows:

(1) Start Aspen Plus and select the default template.

(2) Enter the **Components | Specifications | Selection** page and input components methyl-isobutyl-ketone (MIBK) and dimethyl acetamide (DMAC), as shown in Figure 1.1.

Figure 1.1　Inputting components

(3) Click **Next** to go to the **Methods | Specifications | Global** page and select NRTL method, as shown in Figure 1.2.

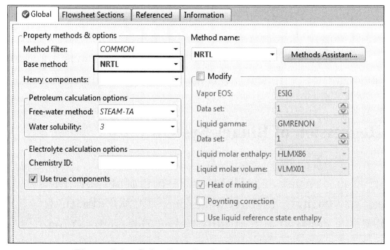

Figure 1.2　Selecting physical property method

(4) Click **Next** to enter **Methods** | **Parameters** | **Binary Interaction** | **NRTL-1** | **Input** page, and find that binary interaction parameters are missing, as shown in Figure 1.3.

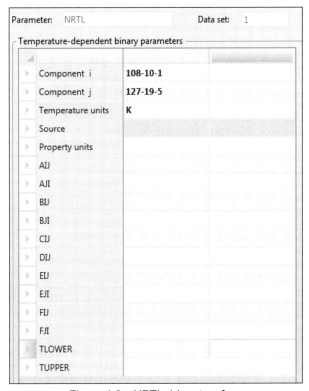

Figure 1.3   NRTL-1 input surface

(5) Enter the Data page, click **New...** button, use the default identification D-1 and select MIXTURE in Select Type, as shown in Figure 1.4.

Figure 1.4   Creating new ID

(6) Click **OK** to enter **Data** | **D-1** | **Setup** page, select TXY in Data type and move MIBK and DMAC from the left column to the right column. Here, the order of component selection is particularly important because MIBK is the first component, as shown in Figure 1.5.

(7) Enter the **Data** | **D-1** | **Data** page and input the experimental data, as shown in Figure 1.6.

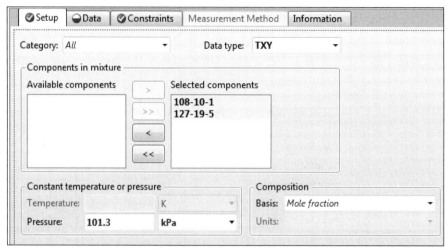

Figure 1.5 Setting the data

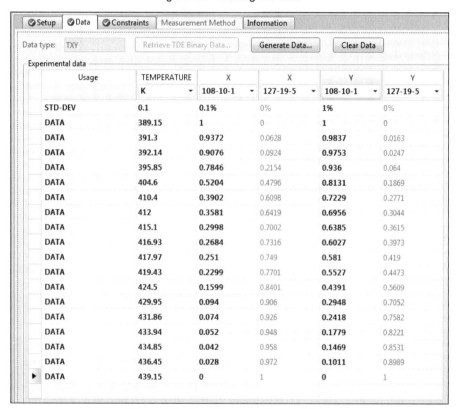

Figure 1.6 Inputting experimental data

(8) Select Run Model in the Home ribbon tab, click **New...** button to establish a new data regression and adopt the default identification R-1, as shown in Figure 1.7.

(9) Click **OK** to enter **Regression | R-1 | Input | Setup** page to set the physical method and data source for regression. Thermodynamic consistency test is performed by default, as shown in Figure 1.8.

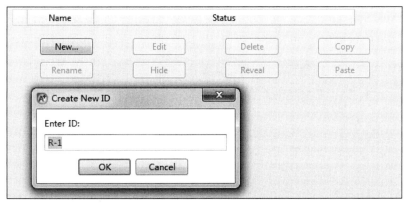

Figure 1.7　Creating new data regression

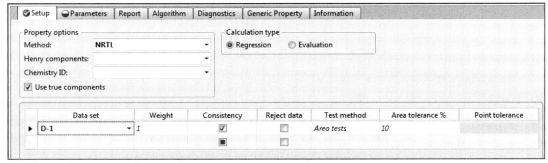

Figure 1.8　Setting data regression

(10) Enter **Regression | R-1 | Input | Parameters** page and set data regression parameters. As shown in Figure 1.9, element 1, 1, 2, 2 and 3 represents the binary interaction parameters of NRTL equation A_{ij}, A_{ji}, B_{ij}, B_{ji} and C_{ij}, respectively.

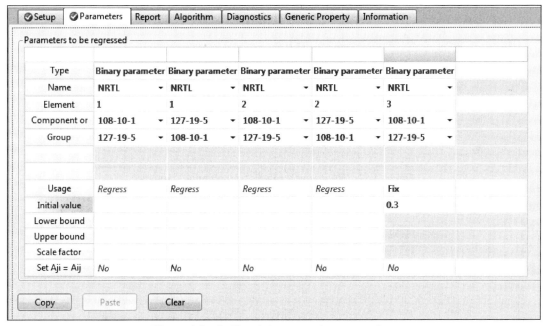

Figure 1.9　Setting data regression parameters

(11) Click **Run** to bring up the **Data Regression Run Selection** dialog box and select the data regression case of R-1 to run, as shown in Figure 1.10.

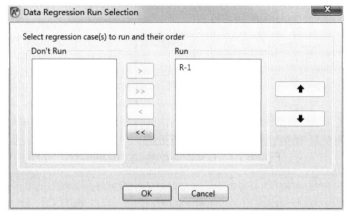

Figure 1.10   Data regression run selection

(12) Click **OK** to run the simulation and the dialog box which is shown in Figure 1.11 will appear. Select **OK**.

Figure 1.11   Running the simulation

(13) Enter **Regression | R-1 | Results | Parameters** page to view regression results and standard deviation, as shown in Figure 1.12.

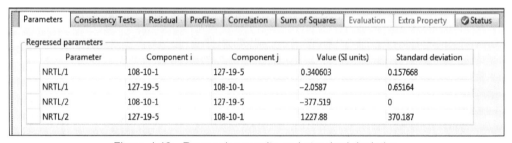

Figure 1.12   Regression results and standard deviation

(14) Enter **Regression | R-1 | Results | Consistency Tests** page and check the thermodynamic test results. It is found that the thermodynamic consistency test has been passed in Figure 1.13.

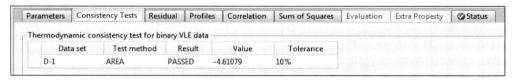

Figure 1.13   The results of Consistency Tests page

(15) Enter **Regression | R-1 | Results | Profiles** page for regression results summary and partial regression results is shown in Figure 1.14.

Exp Val TEMP	Est Val TEMP	Exp Val PRES	Est Val PRES	Exp Val MOLEFRAC X 108-10-1	Est Val MOLEFRAC X 108-10-1	Exp Val MOLEFRAC X 127-19-5	Est Val MOLEFRAC X 127-19-5	Exp Val MOLEFRAC Y 108-10-1
K	K	kPa	kPa					
389.15	389.27	101.3	101.257	1	1	0	0	1
391.3	391.009	101.3	101.4	0.9372	0.937213	0.0628	0.0627866	0.9837
392.14	391.837	101.3	101.406	0.9076	0.907612	0.0924	0.0923876	0.9753
395.85	395.487	101.3	101.43	0.7846	0.784593	0.2154	0.215407	0.936
404.6	404.729	101.3	101.251	0.5204	0.520528	0.4796	0.479472	0.8131
410.4	410.491	101.3	101.264	0.3902	0.390274	0.6098	0.609727	0.7229
412	412.091	101.3	101.264	0.3581	0.35817	0.6419	0.64183	0.6956
415.1	415.23	101.3	101.248	0.2998	0.299879	0.7002	0.700121	0.6385
416.93	417.071	101.3	101.243	0.2684	0.268472	0.7316	0.731528	0.6027
417.97	418.138	101.3	101.233	0.251	0.251071	0.749	0.748929	0.581
419.43	419.506	101.3	101.269	0.2299	0.229932	0.7701	0.770068	0.5527
424.5	424.474	101.3	101.31	0.1599	0.1599	0.8401	0.8401	0.4391
429.95	429.9	101.3	101.32	0.094	0.0939971	0.906	0.906003	0.2948
431.86	431.723	101.3	101.354	0.074	0.0739926	0.926	0.926007	0.2418
433.94	433.805	101.3	101.354	0.052	0.0519948	0.948	0.948005	0.1779
434.85	434.775	101.3	101.33	0.042	0.0419972	0.958	0.958003	0.1469
436.45	436.204	101.3	101.396	0.028	0.0279971	0.972	0.972003	0.1011
439.15	439.098	101.3	101.32	0	0	1	1	0

Figure 1.14 The results of Profiles page

(16) Plot the regression data. Click **Plot | T-xy** to generate the curve (Figure 1.15). It can be seen that the predicted calculation results are consistent with the experimental data, indicating that it is appropriate to fit the two groups of experimental data using NRTL equation.

Figure 1.15 Comparison of experimental data and calculated data

Chapter 1 Simulation of Vapor-liquid and Liquid-liquid Equilibrium for Binary/Ternary Systems

(17) Click **Methods** | **Parameters** | **Binary Interaction** | **NRTL-1** to see the binary interaction parameters in Figure 1.16.

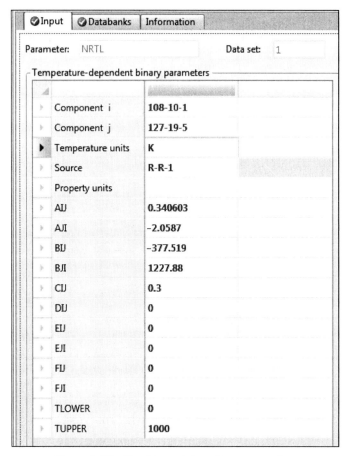

Figure 1.16 The binary interaction parameters

1.3 Data Regression of Ternary System by NRTL

The liquid-liquid equilibrium experiments of the ternary system—methyl tert-butyl ether-methanol-[BMIM] HSO_4 are carried out at 101.3kPa and 298.15K. The binary interaction parameters of the system are obtained by regression of the experimental data with MATLAB 7.0. The NRTL method is selected for physical properties.

二元液液相平衡
实验数据的回归

The simulation steps of this example are as follows:

(1) Open MATLAB 7.0 software, the interface is shown in Figure 1.17.

(2) Click **File** to go to the **Open** page. Choose and open NRTL program that is well edited in advance, as shown in Figure 1.18.

(3) The processed ternary experimental data (mole fraction) of [BMIM] HSO_4

are brought into the pre-programmed NRTL program.

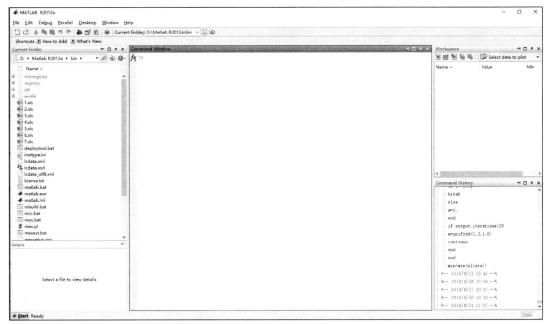

Figure 1.17　MATLAB 7.0 software interface

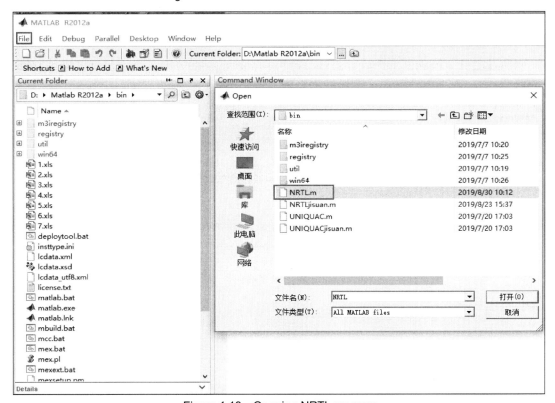

Figure 1.18　Opening NRTL program

(4) Click Start | Toolboxes | More… | Optimization | Optimization Tool (optimtool), as shown in Figure 1.19.

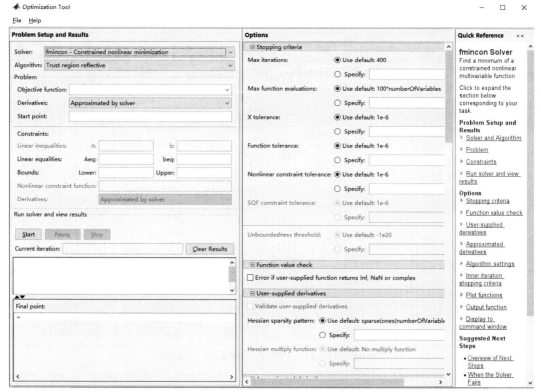

Figure 1.19   Data regression interface

(5) Enter the data regression interface, choose the lsqnonlin – Nonlinear least squares, as shown in Figure 1.20.

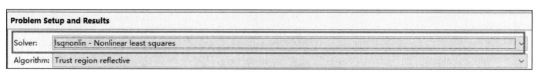

Figure 1.20   Choosing Solver

(6) The objective function is selected to call NRTL program, and input @NRTL. Start point is set the six initial values, as shown in Figure 1.21.

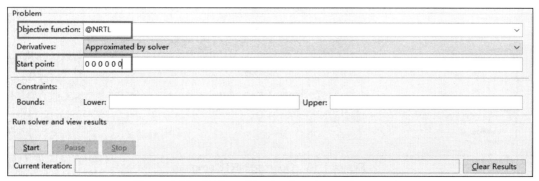

Figure 1.21   Setting objective function and start point

(7) Click **Start** and the program starts running. Current iteration represents

the number of iterations. The default number of iterations for the system is up to 85. The results are shown in the Final point (Figure 1.22).

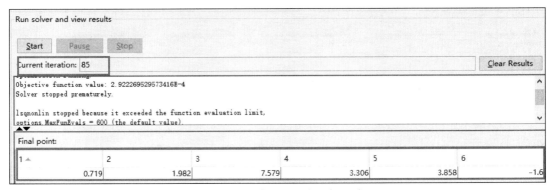

Figure 1.22 Computational results

(8) The binary interaction parameters of component i and j are obtained. The root mean square deviation (RMSD) is calculated, until the results meet the requirements.

1.4 Data Regression of Ternary System by UNIQUAC

The liquid-liquid equilibrium experiments for the ternary system of methyl tert-butyl ether-methanol-[BMIM]HSO$_4$ are carried out at 101.3kPa and 298.15K. The binary interaction parameters of the system are obtained by regression of the experimental data with MATLAB 7.0. The UNIQUAC method is selected for physical properties.

The simulation steps of this example are as follows:
(1) Open MATLAB 7.0 software.
(2) Click **File** to go to the **Open** page. Choose and open UNIQUAC program that is well edited in advance. Repeat step (3)～step (5) mentioned in Chapter 1.3.
(3) The objective function is selected to call UNIQUAC program, and input @UNIQUAC. Start point is set the six initial values, the minimum value is greater than 0, as shown in Figure 1.23.

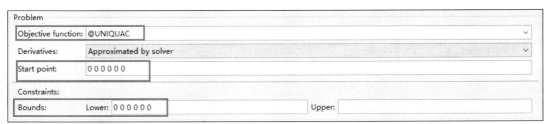

Figure 1.23 Setting objective function and constraints

(4) Click **Start** and the program starts running. The default number of current iterations for the system is up to 85. The results are shown in the Final point (Figure 1.24).

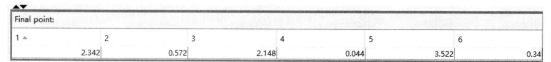

Figure 1.24 Computational results

(5) The binary interaction parameters of component i and j are obtained. The RMSD is calculated, until the results meet the requirements.

Exercises

1. The experimental data of isobaric vapor-liquid equilibrium of ethyl acetate-dimethyl sulfoxide system is shown in Table 1.2. Please regress the experimental data to obtain the binary interaction parameters. The NRTL method is selected for the physical property.

Table 1.2 Experimental data of isobaric vapor-liquid equilibrium for ethyl acetate-dimethyl sulfoxide

T/K	x_1	y_1	T/K	x_1	y_1
350.29	1.0000	1.0000	399.26	0.1322	0.8714
252.94	0.9014	0.9951	414.68	0.0834	0.7759
257.57	0.7124	0.9851	434.61	0.0423	0.5783
381.53	0.2274	0.9373	452.50	0.0153	0.2863
387.82	0.1879	0.9190	463.27	0.0000	0.0000

2. The liquid-liquid equilibrium experiments for the ternary system of methyl tert-butyl ether-methanol-[BMIM] HSO_4 are carried out at 101.3 kPa and 298.15 K. Please regress the experimental data with MATLAB 7.0 to calculate the binary interaction parameters of the system. The liquid-liquid equilibrium experimental data are shown in Table 1.3. The NRTL method is selected for physical properties.

Table 1.3 The liquid-liquid equilibrium experimental data of methyl tert-butyl ether-methanol-[BMIM] HSO_4

Upper phase		Lower phase	
x_1	x_2	x_1	x_2
0.9708	0.0185	0.0095	0.0293
0.9523	0.0437	0.0188	0.0766
0.9231	0.0729	0.0290	0.1345
0.8726	0.1211	0.0534	0.2042
0.8111	0.1731	0.0922	0.2712
0.7424	0.2172	0.1339	0.3178

References

[1] Plus A. Aspen Plus user guide[CP]. United States, 2003.

[2] Wang Y, Zhang H, Pan X, et al. Isobaric vapor-liquid equilibrium of a ternary system of ethyl acetate+ propyl acetate+ dimethyl sulfoxide and binary systems of ethyl acetate+ dimethyl sulfoxide and propyl acetate+ dimethyl sulfoxide at 101.3 kPa[J]. The Journal of Chemical Thermodynamics, 2019, 135: 116-123.

[3] Bai W, Dai Y, Pan X, et al. Liquid-liquid equilibria for azeotropic mixture of methyl tert-butyl ether and methanol with ionic liquids at different temperatures[J]. The Journal of Chemical Thermodynamics, 2019, 132: 76-82.

Chapter 2

Application of Green Solvents in Absorption and Extraction

2.1 Introduction

In recent years, ionic liquid, as a new type of green solvent, has attracted extensive attention of researchers because of its non-volatility, environmentally friendly property, good thermal stability, strong solubility and good designability. This makes it an efficient functional compound in many fields. In the field of separation and purification, ionic liquid shows a strong momentum because of its good solubility to organic and inorganic substances.

However, the high production cost, high viscosity, difficulty in decomposition and potential toxicity of traditional ionic liquids have become the limiting factors for industrial applications. Deep eutectic solvents (DESs), considered as a new type of ionic liquid, have effectively avoided these problems. DESs maintain most favorable properties of ionic liquids and have the advantages of low production cost, low toxicity, good biodegradability and easy synthesis. They can be used as substitutes for organic solvents.

The purpose of this chapter is to regress the experimental data and import it into Aspen Plus for simulating. It reflects the actual situation of the industry and gives some guidance to the industrial application of ionic liquids.

2.2 Molecular Dynamics Simulation

The purpose of this chapter is to guide users of Amber, Amber Tools, and GROMACS to simulate room temperature ionic liquids, calculate their structural properties (RDFs and SDFs) and dynamic

分子动力学模拟实例

properties (self-diffusion coefficients). This simulation method can also help to explore new solvents and thus lead to screen results. However, this calculation method cannot replace experiments, but can supplement them and enrich their applications. This sample tutorial guides new GROMACS users through a simulation, explaining the input files used in each step and the resulting output files. As shown in Figure 2.1, you need to know the general process of molecular dynamics simulation with GROMACS. This tutorial assumes that you are using version 5.0 of GROMACS.

2.2.1 Generating GROMACS Supported Files

With version 5.0 release of GROMACS, all of the tools now become GMX (a module in the program), commands can be used by symbolic links. To get a GROMACS module with the help of the information, you can use any of the following commands GMX help (module) or GMX (module)-h when one of the (module) is used to replace the actual name to query command.

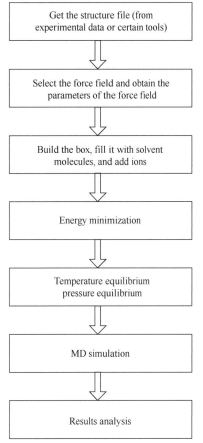

Figure 2.1　GROMACS molecular simulation process

Relevant information will be output to the terminal, including the available algorithms, options, file format, known defects and limitations, and so on. For new users of GROMACS, view the common commands information in help module is an effective way to understand the function of each command.

As shown in Figure 2.2, the entire process of creating small molecule structure GAFF field topology file using AmberTools + ACPYPE + Gaussian is as follows:

The general steps of the creation process are: obtain RESP charge by Gaussian, and then obtain Amber's parameter file by Amber Tools. Since the format of GROMACS and Amber's parameter file is very different, use ACPYPE to convert Amber's parameter file into GROMACS' identifiable gro and top files.

2.2.1.1　*Optimizing Molecular Structure Using Gaussview*

(1) Use Gaussview to create molecular configurations and make preliminary optimization.

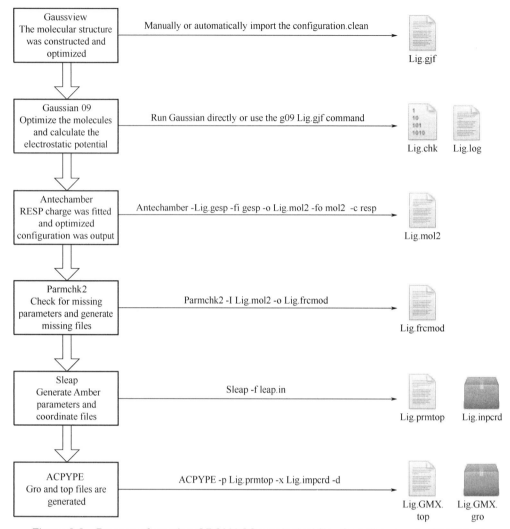

Figure 2.2　Process of creating GROMACS topological file of small molecule GAFF field

Use editing software (eg. Gaussview, Chimera, Chem3D, VMD) to create molecular configurations, as long as the available software and general output support ".pdb" or ".mol2" format. Electrostatic potential needs to be calculated by Gaussian to fit RESP charge. Therefore, it is more convenient to construct molecules and prepare input files with Gaussview.

Once a molecule is built, it is usually necessary to make a rough optimization to make the configuration look more reasonable. This is easy to do in Gaussview by clicking edit-> Clean.

Note: the Gaussian and Gaussview installation path does not contain Chinese, and the input and output files cannot be saved in Chinese.

(2) Edit ".gjf", use a common text editor, such as vi/vim, Emacs (Linux/Mac), or Notepad (Windows). Do not use Word processing software (such as Word under Windows), because they are not suitable for quickly checking the edit plain text

files. If you don't like vi/vim or Emacs mode of operation, you can try VIM or other software. For Windows users, system's own Notepad program is not useful, you can choose other powerful programs, such as Notepad2, Notepad+, EmEditor, EditPlus, UltrEdit, etc. The choice is very wide, choose the one you like, and the one you are most familiar with, so that the processing of various files can be handy.

(3) Then we will submit the ".gjf" files to Gaussian 09 respectively, and use the theory of the B3LYP Gaussian 09 and 6-31 g* basis set to fully optimize all the molecular configurations. The RESP charge fitting method is used to obtain the charge of all molecules. After the normal operation, "Lig. chk", "Lig. out", "Lig_ini. geps" and "Lig. gesp" file will be generated. We only need the "Lig. gesp" file, which is the electrostatic potential file of the optimized configuration, to fit RESP charge, as shown in Figure 2.3.

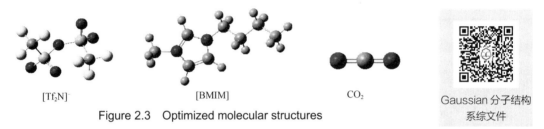

Figure 2.3 Optimized molecular structures

(4) Before proceeding to the next step, install Anaconda 3 and Amber. Anaconda is Python's package manager and environment manager, and it is developed on conda (a package manager and environment manager). There are many third-party packages that you can use for data analysis, and conda is a great way to install, uninstall, and update packages on your computer. So Python needs to be installed to prepare for Amber installation. When Gaussian is done, you get the log file. Use the following command to get the molecular ".mol2" structure file, as shown below.

```
antechamber -i y.log -fi gout -o y.mol2 -fo mol2 -nc -1 -rn ALZ
```

(5) Use Parmchk2 to check GAFF parameters and generate missing parameter files, as shown below.

```
> alz = loadmol2 y.mol2
Loading Mol2 file: ./y.mol2
Reading MOLECULE named ALZ
> check alz
Checking 'alz'....
Checking parameters for unit 'alz'.
Checking for bond parameters.

Error: Could not find bond parameter for: s6 - ne
```

```
Checking for angle parameters.

Error: Could not find angle parameter: s6 - ne - sy

Error: Could not find angle parameter: o - s6 - ne

Error: Could not find angle parameter: o - s6 - ne

Error: Could not find angle parameter: sy - c3 - f

Error: Could not find angle parameter: sy - c3 - f

Error: Could not find angle parameter: sy - c3 - f

Warning: There are missing parameters.
Unit is OK.
```

Parmchk2 is an enhanced version of the original Parmchk, which can check the missing parameters of GAFF in the input molecular configuration and generate the corresponding supplementary parameter file ".frcmod".

2.2.1.2 *Generating Amber Parameter File and Coordinate File Using Tleap*

After ligandopt frcmod is ready, enter the following command in the Tleap command window to load the ligandopt frcmod file into Tleap, as shown below.

```
> loadamberparams y.frcmod
Loading parameters: ./y.frcmod
Reading force field modification type file (frcmod)
Reading title:
Remark line goes here
> check alz
Checking 'alz'....
Checking parameters for unit 'alz'.
Checking for bond parameters.
Checking for angle parameters.
Unit is OK.
```

At this time, save the structure file of ligand small molecule, as shown below. If the saved file is in ".pdb" format, it will not contain the connection information between atoms. When viewing with pymol and other software, it will be found that there is no bond relationship between atoms. If you save a file in ".mol2" format, it will save the key relationship, so it is recommended to save a file in ".mol2" format. At this point, try to save the parameter file and coordinate file that the Amber software needs to use without error reporting.

```
> saveamberparm alz alz.prmtop alz.inpcrd
```
Checking Unit.
Building topology.
Building atom parameters.
Building bond parameters.
Building angle parameters.
Building proper torsion parameters.
Building improper torsion parameters.
 total 0 improper torsions applied
Building H-Bond parameters.
Incorporating Non-Bonded adjustments.
Not Marking per-residue atom chain types.
Marking per-residue atom chain types.
   (Residues lacking connect0/connect1 -
   these don't have chain types marked:

    res total affected

    ALZ 1
)
(no restraints)

2.2.1.3 Converting Amber Files to GROMACS Files Using ACPYPE

As shown below, after obtaining "alz.prmtop" and "alz.inpcrd" files in Amber format, the Python script is used to convert them into GROMACS format:

```
acpype.py -x alz.inpcrd -p alz.prmtop
======================================================================
| ACPYPE: AnteChamber PYthon Parser interfacE v. 0 0 Rev: 0 (c) 2019 AWSdS |
======================================================================
Converting Amber input files to Gromacs ...
==> Writing GROMACS files

Total time of execution: less than a second
```

GROMACS support "lig_gmx.gro", "lig_gmx.top", "em.mdp", "md.mdp", etc. Usually the first two files are needed only.

If you want to process the ".top" file to produce the ".itp" file for inclusion in the protein topology file, you can remove the header, change the atomic type, and remove additional information later.

In fact, the above steps can be completed using ACPYPE, but it is easy to go wrong under Windows due to the path. If you do it step by step, it is easy to locate the specific error step. If you are familiar with the process, you can write these processes automatically in a script, or study how to use ACPYPE to perform successfully.

2.2.2 Defining the Unit Box and Filling Solvent

Now that you are familiar with GROMACS topology files, you can continue to create systems. You can also model proteins or other molecules in different solvents, as long as the species involved have the right field parameters.

Defining a simulation box and adding a solvent are done in two steps:

① Use the editconf module to define the box dimensions, as shown in the code below;

② Insert-molecules modules (called genbox in previous versions) are used to fill the box with water.

Now you need to decide to use what kind of cell. For the purpose of this tutorial, use a simple cubic box as a cell. When you have a better understanding of periodic boundary conditions and box types, use a dodecahedron diamond crystal cell, because its volume is only about 71% of the cubic unit cell at the same periodic distance.

```
zwx@ubuntu:~/Desktop/bmimtf2n top$ gmx editconf -f y.gro -o newbox.gro -c -d 5.0 -bt cubic

Read 15 atoms
Volume: 0.0635233 nm^3, corresponds to roughly 0 electrons
No velocities found
        system size :   0.714   0.368   0.241 (nm)
        diameter    :   0.715                 (nm)
        center      :   0.000  -0.000  -0.001 (nm)
        box vectors :   0.715   0.369   0.241 (nm)
        box angles  :   90.00   90.00   90.00 (degrees)
        box volume  :   0.06                  (nm^3)
        shift       :   5.358   5.358   5.358 (nm)
new center          :   5.358   5.358   5.358 (nm)
new box vectors     :  10.715  10.715  10.715 (nm)
new box angles      :   90.00   90.00   90.00 (degrees)
new box volume      :1230.21                  (nm^3)

Back Off! I just backed up newbox.gro to ./#newbox.gro.1#
```

The above command places [Tf_2N]⁻ in the center of the box (-c), and its distance to the edge of the box is at least 5.0 nm (5.0 -d). The box type is cube (- bt cubic). The distance to the edge of the box is an important parameter. Because we want to use periodic boundary conditions, we must satisfy the minimum mapping convention, that is, an ion can never "see" its own periodic image (not interact with itself), otherwise the calculated force will contain false parts. Insert-molecules modules are used to add ions, as shown below.

```
gmx insert-molecules -f newbox.gro -ci y.gro -o ch_1.gro -nmol 99

gmx insert-molecules -f ch_1.gro -ci ya.gro -o ch_gly.gro -nmol 100

gmx insert-molecules -f ch_gly.gro -ci CO2.gro -o ch_gly_c.gro -nmol 10
```

To generate the ".tpr" file with grompp, an input file with the extension ".mdp" (molecular dynamics parameter) is needed. Grompp will combine the coordinate and topology information with the parameters set in the ".mdp" file to generate the ".tpr" file.

As shown below, the ".mdp" file is usually used for running energy minimization (EM) or molecular dynamics simulation (MD), but in this case it is just used to generate atomic descriptions of the system.

```
; ions.mdp - used as input into grompp to generate ions.tpr
; Parameters describing what to do, when to stop and what to save
integrator      = steep        ; Algorithm (steep = steepest descent minimization)
emtol           = 100.0        ; Stop minimization when the maximum force < 1000.0 kJ/mol/nm
emstep          = 0.01         ; Energy step size
nsteps          = 500000       ; Maximum number of (minimization) steps to perform

; Parameters describing how to find the neighbors of each atom and how to calculate the
interactions
nstlist         = 10           ; Frequency to update the neighbor list and long range forces
cutoff-scheme   = Verlet
ns_type         = grid         ; Method to determine neighbor list (simple, grid)
coulombtype     = PME          ; Treatment of long range electrostatic interactions
rcoulomb        = 1.0          ; Short-range electrostatic cut-off
rvdw            = 1.0          ; Short-range Van der Waals cut-off
pbc             = xyz          ; Periodic Boundary Conditions (yes/no)
```

In fact, any reasonable parameter can be used in the ".mdp" file used in this step. Energy minimization parameter settings are usually used because they are very simple and do not involve any complex combination of parameters. Please note that the files used in this tutorial may only be applicable to opls-aa force fields. Other field settings, especially non-key settings, can be quite different.

As shown below, use the following command to generate the ".tpr" file. It provides an atomic level description of our system. Use the command of gmx genion.

```
gmx grompp -f ions.mdp -c ch_gly_c.gro -p topol.top -o ions.tpr -maxwarn -1
```
Setting the LD random seed to -1430932576
Generated 342378 of the 342378 non-bonded parameter combinations
Generating 1-4 interactions: fudge = 0.5
Generated 342378 of the 342378 1-4 parameter combinations
Excluding 3 bonded neighbours molecule type 'ALZ'
Excluding 3 bonded neighbours molecule type 'BLZ'
Excluding 3 bonded neighbours molecule type 'CLZ'
Removing all charge groups because cutoff-scheme=Verlet
Analysing residue names:
There are: 2100   Other residues
Analysing residues not classified as Protein/DNA/RNA/Water and splitting into groups...
Number of degrees of freedom in T-Coupling group rest is 120897.00

```
gmx genion -s ions.tpr -o ions.gro -p topol.top -neutral yes
```
Reading file ions.tpr, VERSION 2018.4 (single precision)

Neutral guarantees that the total static charge of the system is 0. As shown in Figure 2.4, the molecules instructions from topol ".top" should look like this.

```
[ molecules ]
; Compound              #mols
ALZ                     1000
BLZ                     1000
CLZ                     100
```

Figure 2.4   Molecules instructions from topol .top

2.2.3 Energy Minimization

Now, a neutral system is obtained by adding solvent molecules and ions. Before dynamic simulation, it is necessary to ensure that the structure of the system is normal, not too close to the atomic distance, and the geometric configuration is reasonable. The structure relaxation can meet these requirements, this process is known as energy minimization (EM).

The energy minimization process is similar to the ion addition process. The structure, topology and simulation parameters will be written to the binary input file (tpr) again using grompp, but this time the ".tpr" file is not passed to genion, but the mdrun module of the GROMACS MD engine. As shown below, the input parameter file minim ".mdp" is as follows.

```
; minim.mdp - used as input into grompp to generate em.tpr
integrator   = steep      ; Algorithm (steep = steepest descent minimization)
emtol        = 100.0      ; Stop minimization when the maximum force < 1000.0 kJ/mol/nm
```

```
    emstep       = 0.01       ; Energy step size
    nsteps       = 50000      ; Maximum number of (minimization) steps to perform

    ; Parameters describing how to find the neighbors of each atom and how to calculate the
    interactions
    nstlist        = 10         ; Frequency to update the neighbor list and long range forces
    cutoff-scheme  = Verlet
    ns_type        = grid       ; Method to determine neighbor list (simple, grid)
    coulombtype    = PME        ; Treatment of long range electrostatic interactions
    rcoulomb       = 1.0        ; Short-range electrostatic cut-off
    rvdw           = 1.0        ; Short-range Van der Waals cut-off
    pbc            = xyz        ; Periodic Boundary Conditions (yes/no)
```

As shown below, this parameter file with grompp mdrun gets the binary input file.

```
gmx grompp -f minim.mdp -c ions.gro -p topol.top -o em.tpr
Setting the LD random seed to 1880989698
Generated 342378 of the 342378 non-bonded parameter combinations
Generating 1-4 interactions: fudge = 0.5
Generated 342378 of the 342378 1-4 parameter combinations
Excluding 3 bonded neighbours molecule type 'ALZ'
Excluding 3 bonded neighbours molecule type 'BLZ'
Excluding 3 bonded neighbours molecule type 'CLZ'
Removing all charge groups because cutoff-scheme=Verlet
Analysing residue names:
There are:    2100     Other residues
Analysing residues not classified as Protein/DNA/RNA/Water and splitting into groups...
Number of degrees of freedom in T-Coupling group rest is 120897.00
```

Make sure you've updated the "topol.top" file when you run genbox and genion, otherwise you'll get a bunch of errors (number of coordinates in coordinate file does not match topology, coordinates in coordinate file don't match topology, etc).

Then, call mdrun to minimize the energy: `gmx mdrun -v -deffnm em.`

The -v option is used because it causes mdrun to output more information, which will show the status of each step on the screen. The -deffnm option defines the name of the input file and output file. So, if you don't have the output of grompp, you must explicitly specify its name using mdrun's -s option. For our purposes, we will get the following files.

em.log: ASCⅡ text log file that records the energy minimization process
em.edr: binary energy file
em.trr: full precision binary track file
em.gro: energy minimised structure

Let's do some analysis. Em.edr file contains all the energy terms recorded by GROMACS during energy minimization. You can use GROMACS' energy module to analyze any ".edr" file: gmx energy -f em.edr -o potential.xvg.

When prompted, enter 0 to 10 to select Potential (10) and end with 0 (0). The average value of E_{pot} will be displayed on the screen, and the energy value will be written to the potential ".xvg" file. To use these data for drawing, you can try Xmgrace drawing tools. The results should be similar to those below. It can be seen that E_{pot} converges very well and is stable.

As shown in Figure 2.5, our system is at its lowest energy point, it can be used for real kinetic simulations.

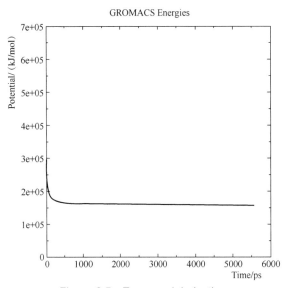

Figure 2.5 Energy minimization

nvt 系综系统文件

2.2.4 NVT Balance

EM guarantees that our initial structure is reasonable in terms of geometry and solvent molecular orientation. In order to simulate the real kinetics, it is necessary to study the solvent and ion equilibrium. If the limit dynamics simulation can be attempted at this time, the system may collapse. The reason is that the solvent molecule itself is basically optimized without considering the solute. So it is necessary to set up a system to determine the reasonable direction of solute under simulated temperature. After reaching the right temperature (based on kinetic energy), it is desirable to exert pressure on the system until the appropriate density is reached.

The purpose of the posre ".itp" file is to apply position restraining forces to heavy atoms (non-hydrogen atoms) in proteins. These atoms do not move unless a very large amount of energy is added. The purpose of positional restriction is that we can balance the solvent molecules around the solute without causing structural changes.

Balance is often divided into two stages. The first stage is carried out under the NVT ensemble (particle number, volume and temperature are constant). This ensemble is also known as the isothermal isometric ensemble or regular ensemble. This process needs time and the construction of the system, but in the NVT ensemble, the temperature of the system should achieve the expected value and remain unchanged. Needed ".mdp" files are shown in the QR code.

Grompp and mdrun will be used, as in energy minimization.

```
gmx grompp -f nvt.mdp -c em.gro -p topol.top -o nvt.tpr
gmx mdrun -deffnm nvt
```

In addition to the comments, a full explanation of the parameters used can be found in the GROMACS manual. Note the following parameters in the ".mdp" file:

Gen_vel = yes: generates initial velocity; using different random number seeds (gen_seed) yields different initial velocity, so multiple (different) simulations can be performed from the same initial structure.

Tcoupl = v-rescale: the velocity rescaling thermostat improves the Berendsen weakly coupled approach, which does not give the correct kinetic energy ensemble.

Pcoupl = no: no pressure coupling is used.

Let's analyze the temperature variation and use the energy module again:

```
gmx energy -f nvt.edr.
```

When prompted, enter 150 to select the system temperature and exit. As shown in Figure 2.6, the result should be similar to the following figure.

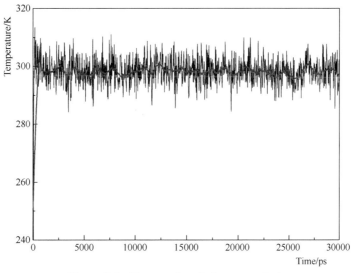

Figure 2.6 Temperature balance analysis

As can be clearly seen from the figure above, the temperature of the system soon reached the target temperature (300 K) and remained stable for the rest of the equilibrium process.

The pressure balance is carried out under the NPT ensemble, which is the second stage of balance, in which the number of particles, pressure and temperature remain stable.

2.2.5 NPT Balance

The NVT equilibrium of the previous step stabilizes the temperature of the system. The pressure (and density) of the system also needs to be stabilized before data can be collected. The pressure equilibrium is carried out under the NPT ensemble, where the number of particles, pressure and temperature remain unchanged.

npt 系综系统文件

This file is not much different from the parameter file used for NVT balance. Note the added pressure coupling section where the parrinello-rahman voltage controller is used.

Grompp and mdrun are used as in NVT equilibrium. Note that we are going to use the -t option to include the NVT balance in the process of the checkpoint file. This file contains all information needed to continue to simulate all state variables. In order to use the speed obtained during the NVT process, we must include this file. The coordinate file (c) is the final output file of NVT simulation.

```
gmx grompp -f npt.mdp -c nvt.gro -t nvt.cpt -p topol.top -o npt.tpr
gmx mdrun -deffnm npt
```

Let's analyze the pressure change and use the energy module again:

```
gmx energy -f npt.edr -o pressure.xvg
```

As shown in Figure 2.7, when prompted, enter 160 to select system pressure and exit. The result should be similar to the following figure.

It is not surprising that the pressure value fluctuates greatly during the equilibrium process. During the whole equilibrium process, the average pressure value is 1.05 bar.

Let's look at density again, using the energy module and typing 220 when prompted.

```
gmx energy -f npt.edr -o density.xvg
```

If you want to draw the average lines in the graph using the cumulative average, it might take a little bit of code to do that, but if you just use the moving average to simply smooth it out, as in the graph, it's pretty easy.

In Xmgrace, in turn, click on the menu **Data -> Transformations -> Running averages.** The length of average should be set according to the characteristics of the specific data. The larger the length, the smoother the average line will be.

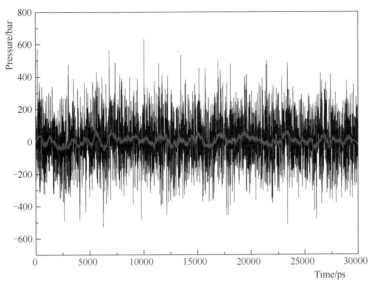

Figure 2.7　Pressure balance analysis

In Origin, click the menu to **analyze -> smooth -> adjacent average** or **FFT filter,** and set the smooth points.

2.2.6　Finishing MD

With two equilibrium phase completed, the system is in need of temperature and pressure balance. It is now possible to liberalize restrictions on product location and MD to collect data. The previous process is similar. Running grompp, the checkpoint files (in this case, including pressure coupling information) must be used, and the parameter files can be download through the QR code.

Run the following commands in turn:

```
gmx grompp -f md.mdp -c npt.gro -t npt.cpt -p topol.top -o md_0_1.tpr
gmx mdrun -deffnm md_0_1
```

md 平衡设置文件

PME load estimation can indicate how many processors should be used for PME calculation, how many processors or PP. Details please refer to the GROMACS related papers and GROMACS manual. For cubic box, the optimal PME load is 0.25, and for dodecahedron box, the optimal PME load is 0.33. When mdrun is executed, the program automatically allocates the number of processors for PP and PME calculations. Therefore, it is necessary to ensure the calculation using the appropriate number of nodes (- the value of np X option). For the system in this tutorial, the computational speed obtained on 12 CPUs is

12 ns/day.

2.2.7 Analysis

Now that the simulation of the systems has been completed, through the analysis of these systems, which types of data are important? This is an important question to consider before simulation, so you should have an idea of which types of data your system needs to collect.

The first module is trjconv, a post-processing tool for processing coordinates, correcting periodicity or manually adjusting tracks (time units, frame frequency, etc). Use the following commands to handle this situation:

```
gmx trjconv -s md_0_1.tpr -f md_0_1.xtc -o md_0_1_noPBC.xtc -pbc mol -ur compact
```
0("System") is selected for the output. According to this "modified" trajectory analysis, first look at the structural stability.

The MSD modules built into GROMACS can be used to calculate RMSD, using the following command to run the tool:

```
gmx rms -s md_0_1.tpr -f md_0_1_noPBC.xtc -o rmsd.xvg -tu ns
```

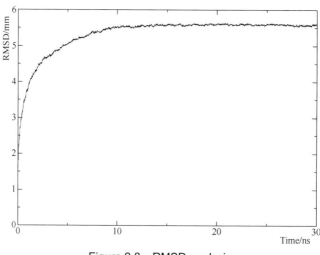

Figure 2.8   RMSD analysis

As shown in Figure 2.8, RMSD indicates a very stable structure. This is expected because it has been energy minimized and, as discussed earlier, location constraints are not 100% perfect.

```
gmx make_ndx -f md_0_1.gro
```

As shown below, create groups for analysis using the following commands:

```
gmx make_ndx -f md_0_1.gro

Reading structure file
Going to read 0 old index file(s)
```

```
Analysing residue names:
There are:    210      Other residues
Analysing residues not classified as Protein/DNA/RNA/Water and splitting into groups...

    0 System              :    4030 atoms
    1 Other               :    4030 atoms
    2 ALZ                 :    1500 atoms
    3 BLZ                 :    2500 atoms
    4 CLZ                 :      30 atoms

  nr : group      '!': not   'name' nr name    'splitch' nr    Enter: list groups
  'a': atom       '&': and   'del' nr          'splitres' nr   'l': list residues
  't': atom type  '|': or    'keep' nr         'splitat' nr    'h': help
  'r': residue               'res' nr          'chain' char
  "name": group              'case': case sensitive           'q': save and quit
  'ri': residue index
```

```
gmx trjconv -s md_0_1.tpr -f md_0_1.xtc -o md_0_1_noPBC.xtc -pbc mol -ur compact
gmx rdf -s md_0_1.tpr -f md_0_1_noPBC.xtc -n index.ndx -o rdf.xvg
```

Use Xmgrace to open the file to view the resulting graph of "rdf.xvg":

```
Xmgrace -nxy rdf.xvg
```

As shown in Figure 2.9, the command above opens the "rdf.xvg" file in Xmgrace, with the X-axis being the distance in angstrom, and $g(r)$ representing the probability of finding CO_2 at a given distance (r) of $[Tf_2N]^-$. The first peak (about four angstrom) represents the first aggregation layer of $[Tf_2N]^-$.

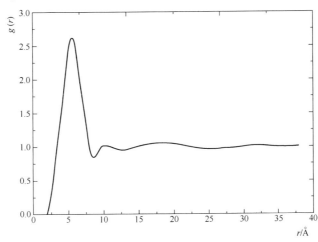

Figure 2.9 Radial distribution function of $[Tf_2N]^-$ and CO_2

According to Einstein's relation, the self-diffusion coefficient D can be calculated by the following formula:

$$D = \frac{1}{6} \lim_{t \to \infty} \frac{\mathrm{d}}{\mathrm{d}t} \left\langle \sum_{i=1}^{N} |r_i(t) - r_i(0)|^2 \right\rangle \qquad (2-1)$$

Where, D represents the self-diffusion coefficient of each component, and the symbol $\langle \rangle$ represents the root-mean-square displacement, namely MSD, and $r_i(t)$ is the coordinate vector representing its center of mass, $r_i(0)$ is the centroid coordinate vector at the initial time.

First, get the mean azimuth shift (MSD) using the following command:

```
gmx msd -f md_0_1.xtc -s md_0_1.tpr -o msd.xvg
```

Xmgrace is used to plot the average MSD slope.

Now that we finish a molecular dynamics simulation with GROMACS, and analyze the results. You can also complete more types of simulation with GROMACS (free energy calculation, non-equilibrium MD, normal mode analysis, etc). You should read some literatures and GROMACS manuals, try to adjust the parameters in the ".mdp" file provided here to make the simulation more efficient and accurate.

2.3 Simulation of Extractive Distillation Using the Ionic Liquid

NRTL or UNIQUAC models can be selected as global physical methods for simulation of ionic liquids (ILs) extractive distillation. This requires defining ionic liquids in Aspen Plus, besides the physical parameters such as viscosity, enthalpy and critical parameters of ionic liquids, binary interaction parameters between ionic liquids and other components are also required. The parameters of binary interaction can be obtained by referring to relevant literatures and the experimental data of liquid-liquid equilibrium.

2.3.1 Analysis of Correlation Model

The liquid-liquid equilibrium data and ternary phase diagrams between ILs and mixtures can be obtained through the liquid-liquid extraction experiments of ILs. The reliability of the experimental data can be tested by Othmer-Tobias equation, Bachman equation and Hand equation. The expressions of the above equations are as follows:

$$\ln\left(\frac{1-x_1^{\mathrm{I}}}{x_1^{\mathrm{I}}}\right) = a + b\ln\left(\frac{1-x_3^{\mathrm{II}}}{x_3^{\mathrm{II}}}\right) \tag{2-2}$$

$$x_3^{\mathrm{II}} = a + b\left(\frac{x_3^{\mathrm{II}}}{x_1^{\mathrm{I}}}\right) \tag{2-3}$$

$$\ln\left(\frac{x_2^{\mathrm{II}}}{x_3^{\mathrm{II}}}\right) = a\ln\left(\frac{x_2^{\mathrm{I}}}{x_1^{\mathrm{I}}}\right) + b \tag{2-4}$$

Among them, x_1^{I} is the mole fraction of solvent in aqueous phase, x_3^{II} is the mole fraction of ionic liquid in extraction phase, x_2^{I} is the mole fraction of extracted substance in extraction residual phase, x_2^{II} is the mole fraction of extracted substance in ionic liquid phase, a and b are constant. The correlation coefficient and fitting degree R^2 of the equation are obtained by fitting the above equation. The closer the correlation coefficient of the data fitting is set to 1, the higher the reliability of the data is.

NRTL equation was proposed by Renon et al in 1968. It can be used to calculate the activity coefficient of non-ideal system. The expression of NRTL equation is as follows:

$$\ln \gamma_i = \frac{\sum_j x_j \tau_{ji} G_{ji}}{\sum_k x_k G_{ki}} + \sum_j \frac{x_j G_{ij}}{\sum_k x_k G_{kj}} \left(\tau_{ij} - \frac{\sum_m x_m \tau_{mj} G_{mj}}{\sum_k x_k G_{kj}} \right) \qquad (2\text{-}5)$$

$$G_{ij} = \exp(-\alpha_{ij} \tau_{ij}) \qquad (2\text{-}6)$$

$$\tau_{ij} = \frac{\Delta g_{ij}}{RT} \qquad (2\text{-}7)$$

Among them, the formula $\tau_{ii}=0$, $G_{ii}=1$ are used for calculating the excess free energy G^{E}, then NRTL equation becomes:

$$\frac{G^{\text{E}}}{RT} = \sum_i x_i \frac{\sum_j x_j \tau_{ji} G_{ji}}{\sum_k x_k G_{ki}} \qquad (2\text{-}8)$$

UNIQUAC model was proposed by Abrums et al in 1975. The expression of multivariate activity coefficient (γ) equation of UNIQUAC model is as follows:

$$\ln \gamma_i = \ln \frac{\Phi_i}{x_i} + \frac{z}{2} q_i \ln \frac{\theta_i}{\Phi_i} + l_i - \frac{\Phi_i}{x_i} \sum_j x_j l_j + q_i' \left[1 - \ln \left(\sum_j \theta_j' \tau_{ji} \right) - \sum_j \frac{\theta_j' \tau_{ij}}{\sum_k \theta_k' \tau_{kj}} \right] \qquad (2\text{-}9)$$

$$\theta_i = \frac{q_i x_i}{\sum_j q_j x_j} \qquad (2\text{-}10)$$

$$\theta_i' = \frac{q_i' x_i}{\sum_j q_j' x_j} \qquad (2\text{-}11)$$

$$\Phi_i = \frac{r_i x_i}{\sum_j r_j x_j} \qquad (2\text{-}12)$$

$$l_i = \frac{z}{2}(r_i - q_i) - (r_i - 1) \qquad (2\text{-}13)$$

For most substances, except for water and some small alcohols, $q_i = q_i'$. z means coordination number, generally $z = 10$. τ_{ij} and τ_{ji} are adjustable parameters between the two binary substances, and they are expressed as follows:

$$\tau_{ij} = \exp\left[-\frac{\Delta u_{ij}}{RT}\right] \quad (2\text{-}14)$$

$$\tau_{ji} = \exp\left[-\frac{\Delta u_{ji}}{RT}\right] \quad (2\text{-}15)$$

Δu_{ij} and Δu_{ji} are the interaction parameters between component i and j.

The program is conducted with MATLAB, and the experimental data are correlated with non-linear least squares method in the software. The objective function is iterated by minimizing the sum of squares of deviations between the experimental and calculated values. The expression of the objective function is as follows:

$$\text{OF} = \sum_{k=1}^{M}\sum_{j=1}^{2}\sum_{i=1}^{3}\left(x_{ijk}^{\exp} - x_{ijk}^{\text{calc}}\right)^2 \quad (2\text{-}16)$$

Among them, M is the number of connecting lines, $x^{\exp}$ is the experimental value of component mole fraction, x^{calc} is the calculated value of component mole fraction, i represents component, j is the phase number, k is the connecting line number. In the regression process, the value of the objective function is set to 10^{-6}.

The root mean square deviation (RMSD) is used to test whether the binary interaction parameters obtained are suitable for the research system. The expression is as follows:

$$\text{RMSD} = \left(\sum_{k=1}^{M}\sum_{j=1}^{2}\sum_{i=1}^{3}\frac{\left(x_{ijk}^{\exp} - x_{ijk}^{\text{calc}}\right)^2}{6M}\right)^{1/2} \quad (2\text{-}17)$$

The basic equation of liquid-liquid equilibrium used in data regression is as follows:

$$\gamma_i^{\text{I}} x_i^{\text{I}} = \gamma_i^{\text{II}} x_i^{\text{II}} \quad (2\text{-}18)$$

Among them, x_i^{I} is the mole fraction of component i in the extraction residue at equilibrium, x_i^{II} is the mole fraction of component i in the extraction phase at equilibrium, γ_i^{I} is the activity coefficient of component i in the extraction residue, and γ_i^{II} is the activity coefficient of component i in the extraction phase. In UNIQUAC model, molecular volume parameter r and area parameter q are needed. The values of r and q of common substances can be found in relevant literature and database, but the values of r and q of ionic liquids need to be calculated or consulted.

2.3.2 Definition of the Ionic Liquid in Aspen Plus

In this chapter, [Emim][BF$_4$] is used to separate the ethanol-water system.

(1) Open Aspen Plus V10 and create a new blank simulation, all the compounds should be input into Aspen Plus, click **User Defined** to input the name of the ionic liquid, as shown in Figure 2.10.

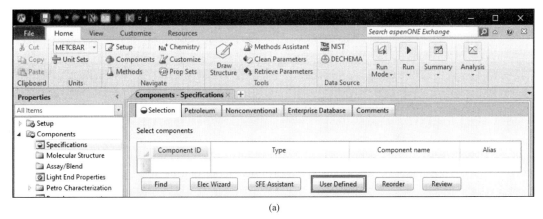

(a)

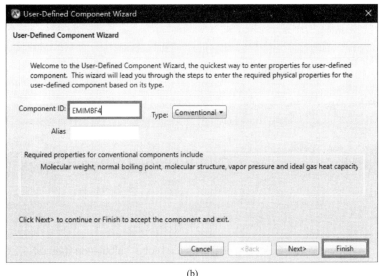

(b)

Figure 2.10　Inputting ionic liquid

(2) Use UNIFAC group contribution method to simply define the ionic liquid in Aspen Plus to prevent Aspen from reporting errors due to the missing structure of [Emim][BF$_4$]. Click **Components | UNIFAC Groups** and define the UNIFAC groups, here the number of ionic liquid is set to 4000, as shown in Figure 2.11.

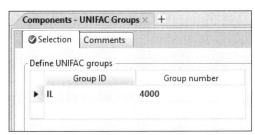

Figure 2.11　Defining UNIFAC groups

Click **Components| Molecular Structure | DES-CU | Functional Group** and **UNIFAC**, input the UNIFAC groups of [Emim][BF$_4$], as shown in Figure 2.12.

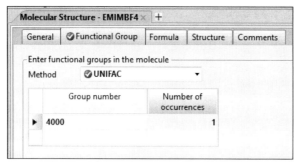

Figure 2.12 Inputting the UNIFAC groups of [Emim][BF$_4$]

(3) NRTL is chosen as the global physical method, as shown in Figure 2.13.

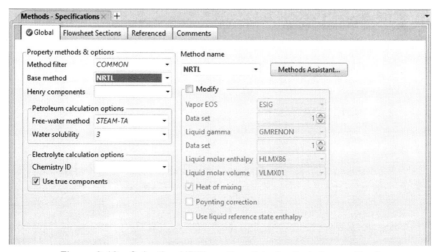

Figure 2.13 Selecting NRTL as the global calculation method

(4) Click **Parameters** to enter Pure Components page and press **New**, as shown in Figure 2.14.

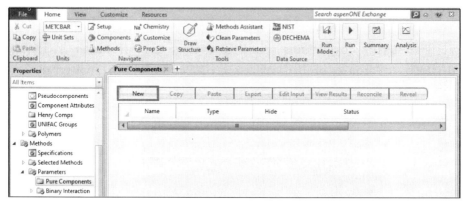

Figure 2.14 Setting pure component parameter

Select Scalar as the type of pure component parameter and click **OK**, then input the molecular weight, bubble point, critical properties, eccentricity factor of [Emim][BF$_4$], as shown in Figure 2.15.

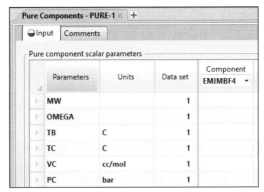

Figure 2.15 Inputting the physical properties of the component

Once again, select T-dependent correlation | Liquid heat capacity | CPLPO-1. Input the ionic liquid ideal gas heat capacity equation coefficient, as shown in Figure 2.16.

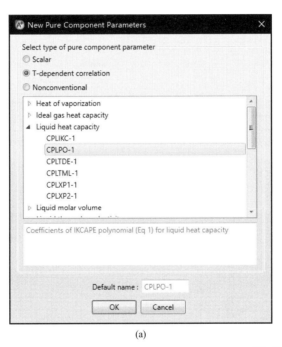

(a)

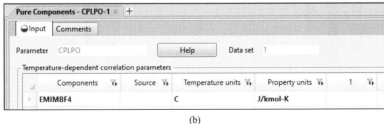

(b)

Figure 2.16 (a) Selecting CPLPO-1; (b) Inputting the ionic liquid ideal gas heat capacity equation coefficient

Select T-dependent correlation | Liquid viscosity | MULAND-1 and input the Andrade liquid viscosity equation coefficient, as shown in Figure 2.17.

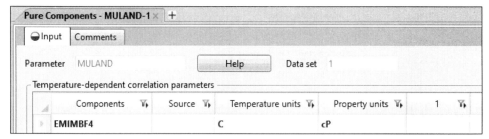

Figure 2.17 Inputting the Andrade liquid viscosity equation coefficient

(5) Input the missing binary interaction parameters in **Methods | Parameters | Binary Interaction**, as shown in Figure 2.18.

Click **Run**, and Aspen will calculate some properties of the ionic liquid, ionic liquids are defined.

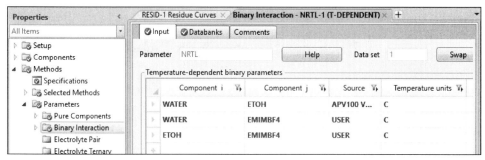

Figure 2.18 Inputting the missing binary interaction parameters

(6) Click **NEXT** to enter the simulation environment. The number of extractive distillation trays is 40, and the mixture enters from the 34th tray. IL enters the tower from the 7th tray as the solvent. The pure ethanol is ejected and the bottom stream passes to a conventional rectification column to separate water from the ionic liquid. The feed composition and process information are shown in Figure 2.19.

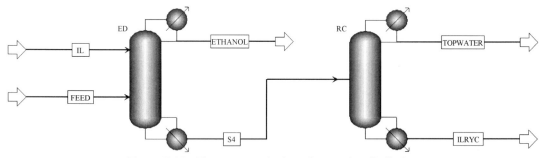

Figure 2.19 The process design of extractive distillation

The extractive distillation column is set to condense in the whole column, and the convergence method is strongly non-ideal liquid, as shown in Figure 2.20.

The stream from the extractive distillation column enters the rectification column from the 12th tray, and the distillation column parameters are set as follows (Figure 2.21).

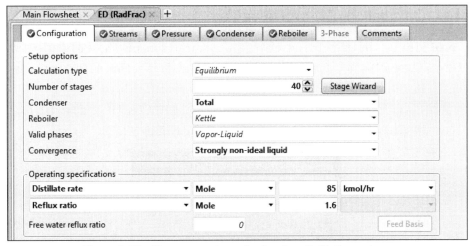

Figure 2.20 The parameters of extractive distillation column

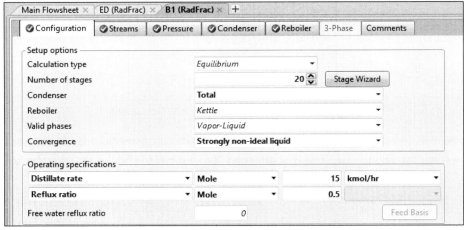

Figure 2.21 The parameters of rectification column

After the separation process is conducted, the ethanol concentration at the top of the column reaches 0.999 (mole fraction), and the IL concentration is 0.9999 (mole fraction). The simulation results can be viewed in the Results Summary.

2.4 Simulation of CO_2 Absorption Using the Ionic Liquid

The vast of ionic liquids do not exist in Aspen Plus databases and need to be defined by users. The key of selecting ionic liquids is the characteristic parameters of the ionic liquid, such as viscosity parameters, enthalpy parameters and critical parameters, which need to be input into Aspen Plus. The selection of physical method is important for the whole simulation

运用 Materials Studio 计算离子液体 σ-profile

process. Here, the COSMO-SAC model is chosen as the global physical method.

2.4.1 Calculation of σ-profile Value

σ-profile is the shielding charge distribution of the molecule, which is input into Aspen Plus as the main parameter and it is the main property of the component. Materials Studio software is used to calculate it.

(1) Open Materials Studio software and select **Create a new project**. Select 3D Atomistic Document, as shown in Figure 2.22.

Figure 2.22 Selecting 3D Atomistic Document

Draw a model of the structure of the bat for the molecule needed, take water as an example. Then, use the Clean tool to correct the bond lengths and bond angles.

(2) Now, move onto setting up the geometry optimization calculation. Select the DMol3 calculation from the Modules menu or use the icon on the toolbar, as shown in Figure 2.23.

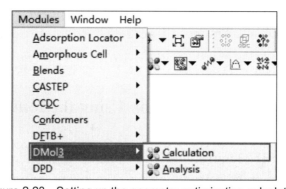

Figure 2.23 Setting up the geometry optimization calculation

Select the task option in the DMol3 calculation. Geometry optimization fixed the atomic coordinates for the molecule by minimizing the total energy of the molecule.

Select GGA (local correlation) and VWN-BP (gradient-corrected function) for the functional option in the Setup tab, as shown in Figure 2.24.
- GGA = Generalized Gradient Approximation.
- VWN-BP = BP function with local correlation replaced by the VWN function.

Koch and Holthausen (2001) provide excellent guidance in choosing appropriate functionality.

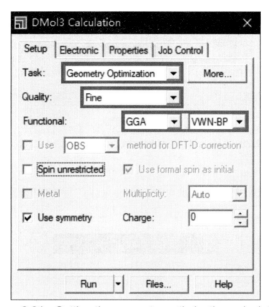

Figure 2.24 Setting the geometry optimization calculation

Adjust the geometry optimization tolerance to fine. It can be done in the Setup or Electronic tab. This is the accuracy of the Hamiltonian matrix element convergence.

This specifies the accuracy to which the Self-Consistent Field (SCF) equations are converged. The fine setting is recommended by the software documentation for highly accurate geometry optimization. It represents a convergence of 10^{-6}.

Select DNP as the basis set in the Electronic tab.

DNP = Double numerical basis with polarization functions. Functions with angular momentum is higher than the highest occupied orbital in the free atom.

According to the software documentation, minimal basis sets are generally inadequate for anything except qualitative results, while DNP sets are the most reliable. The DNP option instructs the program to ignore extraneous functions that "eliminate" certain atomic orbitals.

When quality is set to fine, DNP is the default basis set, as shown in Figure 2.25.

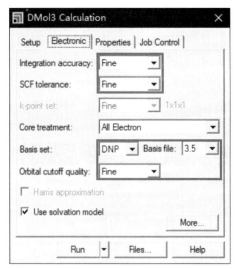

Figure 2.25 The setting of Electronic tab for DMol3 calculation

Click the **More...** button in the Job Control tab.

Check the **Retain server files** box. This will save the files to the hard drive in the jobs directory, as shown in Figure 2.26.

Figure 2.26 Selecting Retain server files

Click **Files...** at the bottom of the DMol3 Calculation dialog.

Click **Save Files** and create the input files necessarily for the calculation, as shown in Figure 2.27. They are visible in the Project window.

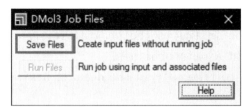

Figure 2.27 Saving files for geometry optimization

Open H_2O input file from the Project window and insert the code "Basis_version

V4.0.0" into the input file. This statement instructs the program to use V4.0.0 DNP basis set instead of the default V3.5 DNP basis set. The software documentation recommends V4.0.0 for COSMO calculations, as shown in Figure 2.28.

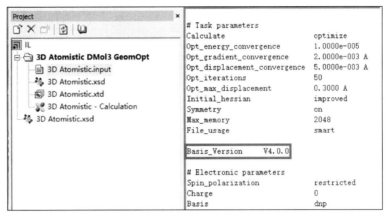

Figure 2.28 Modifying H_2O input file

Keep the Spin unrestricted box unchecked in the Setup tab.

Check the Spin unrestricted box when working with radicals, charged molecules and organometallics. Although it is possible to run an organic molecule with the spin unrestricted setting, it will be much slower computationally.

Check the Use symmetry box in the Setup tab and apply it to certain molecules.

Click **Run Files** to begin the DMol3 geometry optimization, as shown in Figure 2.29.

The geometry optimization predicts the energy level of the molecule in the ideal gas phase.

This calculation is the longest step of the procedure and will require 75% of the time to produce a σ-profile.

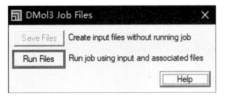

Figure 2.29 Running files

The Project window will show the files which are created and tracked by the progress of the calculation. The ".xsd" file is the optimized geometry. All further calculations should use this file. The ".outmol" file will show the ideal gas phase energy. There are additional files that are not shown in the Project window but they are stored in the Jobs folder.

Geometry optimization has been completed. Proceed to the COSMO calculation.

(3) Open the geometry optimization output, vt-0003.xsd and open another DMol3 Calculation dialog box. Select the task Energy. Leave all other options as same

as geometry optimization. Click **Files...** in the Calculation window.

Click the **Save Files** button to create the input files.

Open the H₂O input file and add the COSMO calculation keywords.

Now review the parameters and settings in the COSMO keywords. This statement turns the COSMO calculation. Again, the base set V4.0.0 will be used instead of the default V3.5. The first number is the atomic number and the second term is the fitted atomic radii. Klamt (1998) fitted these parameters to several sets of data.

Run the COSMO calculation. Click **Run Files** from the Calculation window to begin the energy calculation.

Upon completion, the Project window will contain ".cosmo" and ".outmol" files. There are additional files storing in the Jobs folder. The ".cosmo" file contains many useful information. This file contains the volume of the cavity around the molecule in them theoretical conducting medium, and it is used in the COSMO-RS/COSMOSAC models. It also contains the condensed phase energy, the number of surface segments, and their charge. The ".cosmo" file is edited to make it compatible with the sigma averaging FORTRAN program. Record the number of surface segments.

Delete all information above the red line, which contains the segment charges and coordinates. Keep segment information only, as shown in Figure 2.30.

Then save it as a ".txt" file.

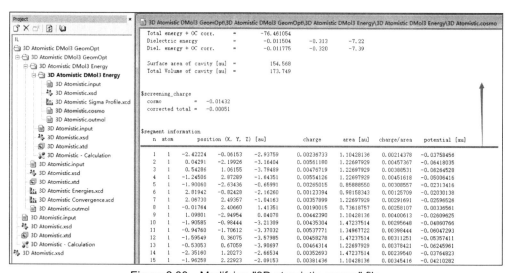

Figure 2.30　Modifying "3D atomistic.cosmo" file

(4) Get the program FORTRAN that calculate σ-profile in www.design.che.vt.edu/vt-2005.html. The output file of the FORTRAN program is located in the C:\Profiles\ directory. This directory must be created manually before running the averaging program. The output file can be input directly from the COSMO-SAC program or can be plotted graphically. We run the FORTRAN program to average the

surface charges over a standardized bonding site.

Run sigma profile.exe, as shown in Figure 2.31.

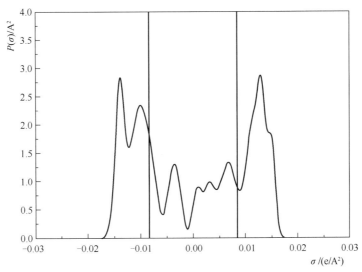

Figure 2.31　Using the software to calculate σ-profile

The type of sigma averaging program output file is text, it can be easily converted into graphical form, as shown in Figure 2.32.

Figure 2.32　The σ-profile of H_2O

The calculation process of ionic liquids is the same as that of the preceding example, except that the σ-profile values calculated from anions and cations are added up.

2.4.2　Definition of the Ionic Liquid in Aspen Plus

In this section, the CO_2 absorption process using ionic liquid [BMIM][Tf_2N] is the solvent simulated.

The definition of ionic liquid is described in section 2.3.2, but there is some deviation because of the CSOMO-SAC model.

(1) COSMOSAC is selected as physical property method, as shown in Figure 2.33.

(2) At the PURE-1 page, the volume parameters of all compounds in COSMO-SAC model need to be added, as shown in Figure 2.34.

运用COSMO-SAC的方法在Aspen Plus中定义离子液体

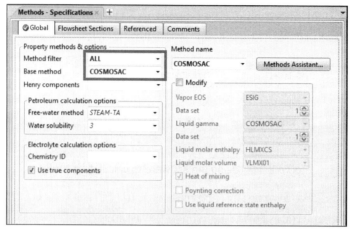

Figure 2.33　Selecting COMSOSAC as the global calculation method

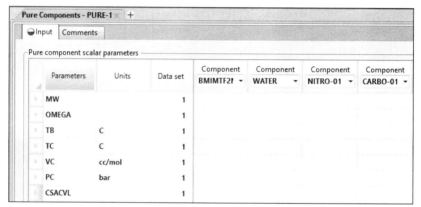

Figure 2.34　Inputting the physical properties of components

(3) The σ-profile parameters of each component in COSMO-SAC model need to be input. Select **T-dependent correlation** | **CSOMO-SAC** | **SGPRF** and input all component's molecular σ-profile parameters, as shown in Figure 2.35.

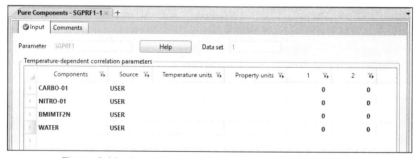

Figure 2.35　Inputting all component's molecular σ-profile

Click **Run**, and Aspen will calculate some properties of the ionic liquid. The definition of ionic liquid is completed.

2.4.3　Simulation of CO_2 Capture Using the Ionic Liquid

Carbon capture using ionic liquid as solvent is considered as an effective way to reduce CO_2 emission. The aim of this chapter is to put forward an ionic liquid (IL)-based CO_2 capture process to achieve the CO_2 capture from the power plants flue gas. Ionic liquid 1-butyl-3-methylimidzolium bis (trifluoro methane sulfonyl, [BMIM][Tf$_2$N]) is used as the solvent. The composition of mixed gas is CO_2 and N_2, as shown in Figure 2.36.

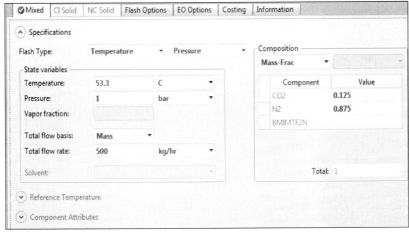

Figure 2.36 The composition of the feed stream

The flowsheet is exhibited in Figure 2.37.

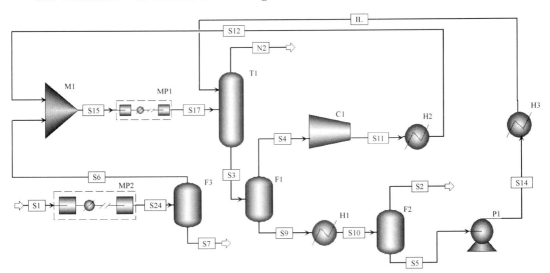

Figure 2.37 The process design of CO_2 absorption

This process mainly consists two section, including absorption and adsorption. The mixed gas is pressurized through multistage compressor (M-CP1). The parameters of multistage compressor are shown in Figure 2.38.

[BMIM][Tf$_2$N] has a good absorption effect of CO_2 under high pressure. The mixed gas and IL enter absorber to achieve the absorption of CO_2. N_2 is separated from the absorber, then the CO_2-enriched solvent enters flash-2. The detailed performance parameters of absorber are shown in Figure 2.39.

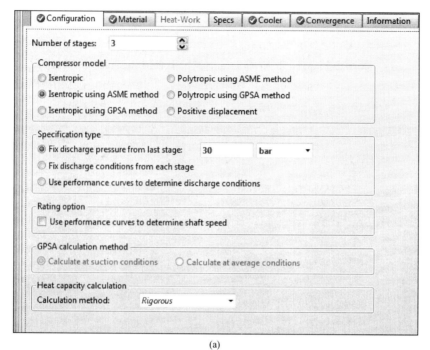

(a)

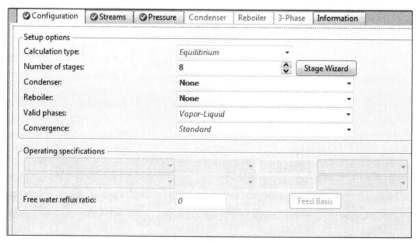

(b)

Figure 2.38 The performance parameters of multistage compressor

(a)

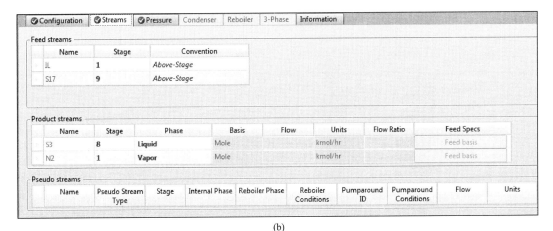

(b)

(c)

Figure 2.39　The performance parameters of absorber

It is noteworthy that the "no" should be set to "yes" for the absorption tower as shown in Figure 2.40.

The gas released from flash-1 is recirculated to the absorber and the CO_2-enriched solvent is sent to flash-2 to achieve the CO_2 capture and solvent recovery. The detailed performance parameters of flash-1 are shown in Figure 2.41 and flash-2 are shown in Figure 2.42.

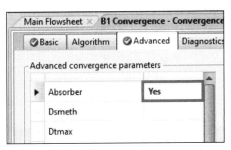

Figure 2.40　The absorber parameter

The IL flows through pumps and heat exchanger, then enters the first absorption column for continuous absorption. The detail is shown in Figure 2.43.

Chapter 2　Application of Green Solvents in Absorption and Extraction

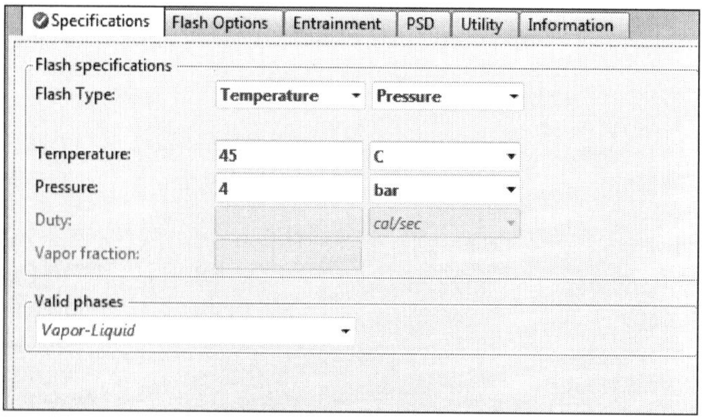

Figure 2.41　The detailed performance parameters of flash-1

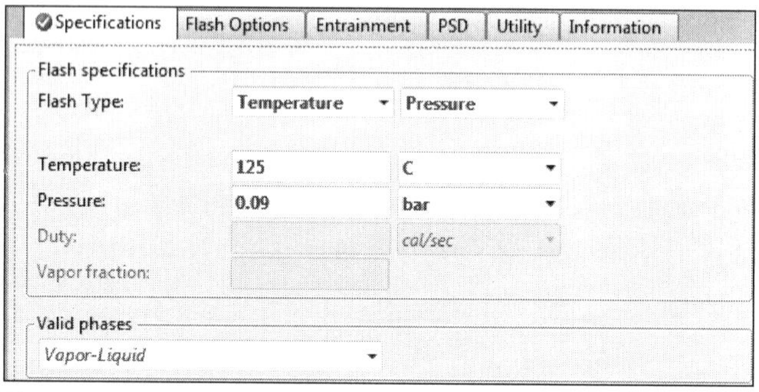

Figure 2.42　The detailed performance parameters of flash-2

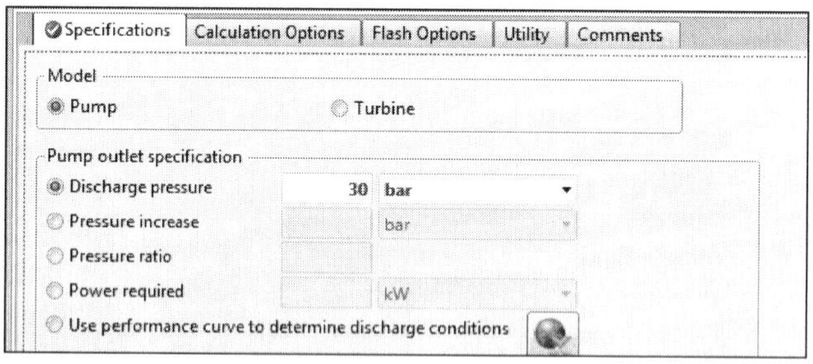

(a)

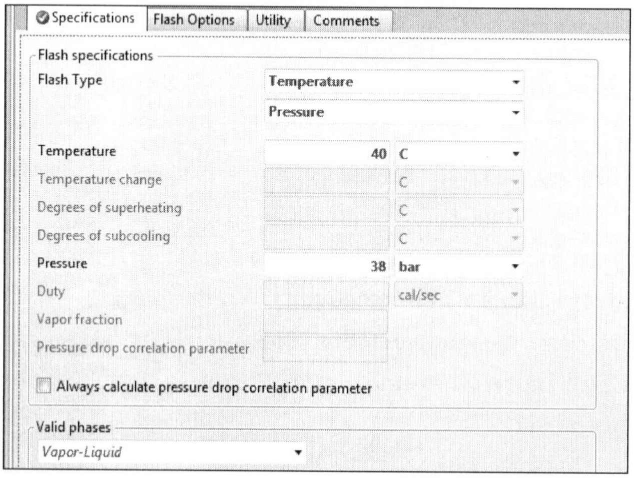

(b)

Figure 2.43　The performance parameters of pumps and heat exchanger

The simulation results can be viewed in Figure 2.44 and Figure 2.45.

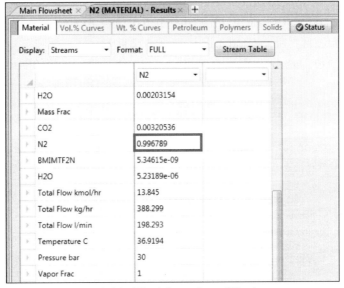

Figure 2.44 The information of N_2 stream

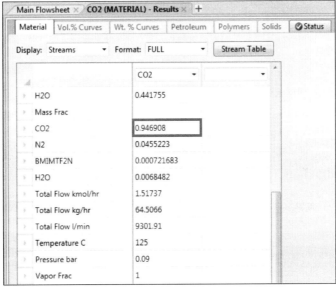

Figure 2.45 The information of CO_2 stream

2.5 Simulation of Extractive Distillation Using Deep Eutectic Solvents

Deep Eutectic Solvents (DESs) refer to a two-component or three-component eutectic mixture composed of hydrogen bond acceptor and hydrogen bond donor with a certain stoichiometric ratio, which solidification point is significantly lower than the melting point of pure substances of each component.

The physical and chemical properties of DESs are similar to those of ionic liquids, so it is also classified as a new type of ionic liquids or ionic liquids analogues. Compared with ionic liquids, DESs have attracted much attention due to their low toxicity and good biodegradability.

In this chapter, we briefly describe the simulation of DESs (ChCl:Urea=1:2) as a solvent for the extraction of ethanol-water binary azeotrope.

2.5.1 Definition of Deep Eutectic Solvents in Aspen Plus

Input each component in Aspen Plus and name the selected DES, chloride/urea (1∶2 on a molar basis) was chosen as solvent to separate ethanol-water binary system, as shown in Figure 2.46.

Figure 2.46 Inputting each component into Aspen Plus

Using UNIFAC group contribution method simply defines DES into Aspen Plus. First define the group, click **Components | UNIFAC Groups**, here we define the UNIFAC group representing the DES structure as 4000, as shown in Figure 2.47.

Click **Components | Molecular Structure | DES-CU | Functional Group** and choose **UNIFAC** as method, input the UNIFAC groups of the DES, as shown in Figure 2.48.

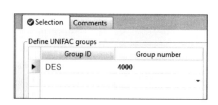 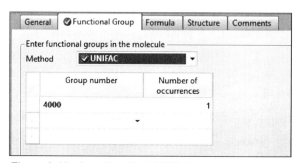

Figure 2.47 Defining UNIFAC Groups   Figure 2.48 Inputting the UNIFAC groups of the DES

The physical method is NRTL.

As shown in Figure 2.49, input the properties of the DES used in the Aspen Plus simulation. Click **Methods | Parameters | Pure Components | NEW**, choose **Scalar**, and input molecular weight, eccentricity factor, critical properties, etc.

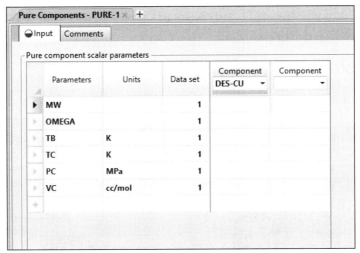

Figure 2.49 Inputting the properties of the DES

As shown in Figure 2.50, click **T-dependent correlation | Ideal gas heat capacity | CPIG-1**, input the coefficients for the ideal gas heat capacity equation of the DES.

Figure 2.50 Inputting the coefficients for the ideal gas heat capacity equation

Click **T-dependent correlation | Liquid viscosity | MULAND-1**, input the coefficients of IKCAPE polynomial for liquid viscosity of the DES.

Click **Methods | Parameters | Binary Interaction | NRTL-1**, input the binary interaction parameters of NRTL model, the binary interaction parameters are obtained by the regression of actual experimental data.

Click **Run**, Aspen Plus will automatically calculate some missing data, the DES definition is completed.

2.5.2 Process Simulation

This simulation is the extraction of an ethanol-water binary azeotrope using DES. The mixed stream feeds from the bottom of the extractive distillation column, and the solvent DES feeds from the top. After separation, high-purity ethanol is obtained from the top of the column, as shown in Figure 2.51.

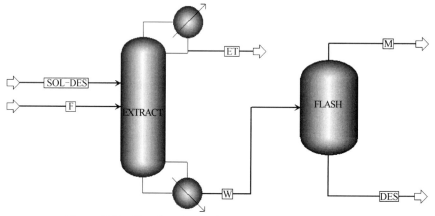

Figure 2.51 The flowsheet of extractive distillation process

Feed parameters are shown in Figure 2.52 and Figure 2.53.

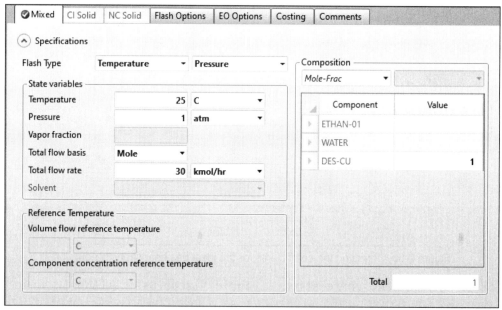

Figure 2.52 The mixture feed information

The parameter of distillation column is set as shown in Figure 2.54. Flash tower parameter is set as shown in Figure 2.55.

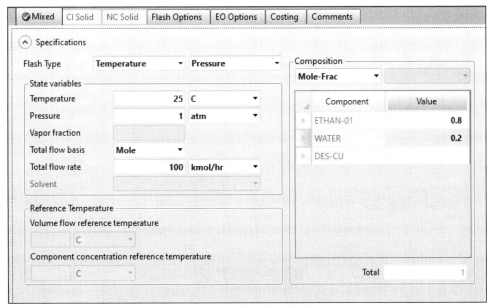

Figure 2.53 The DES feed information

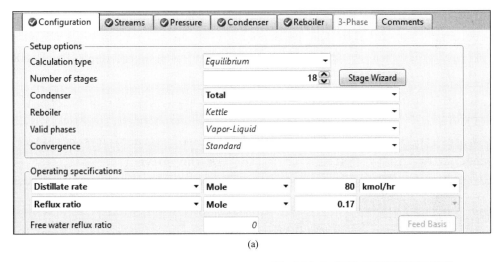

(a)

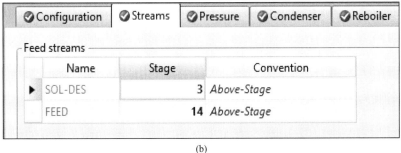

(b)

Figure 2.54 Distillation column parameter setting and feed location

Click **Run**, run the simulation file, the simulation results can be viewed in **Results Summary | Streams**.

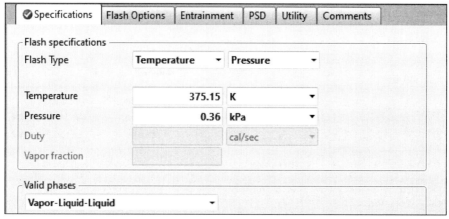

Figure 2.55　Setting flash tower parameters

Exercises

1. Please complete the molecular dynamics simulation of 1-butyl-3-methylimidazole tetrafluoroborate ([BMIM][BF_4]) and acetonitrile (CH_3CN) to reproduce the data in the reference. (The case is from Separation and Purification Technology, 2019, 219: 113-126 and Phys. Chem. Chem. Phys, 2005, 7 (14): 2771-2779.)

2. Calculate the σ-profile value of [Emim][BF_4] using MS and COSMO software.

3. Try to use COSMO-SAC model to simulate extraction separation of ethanol-water binary system, [Emim][BF_4] is the solvent.

References

[1] Heintz Y J, Sehabiague L, Morsi B I. Hydrogen sulfide and carbon dioxide removal from dry fuel gas streams using an ionic liquid as a physical solvent[J]. Energy & Fuels, 2009, 23(10): 4822-4830.

[2] Jalili A H, Shokouhi M, Maurer G. Solubility of CO_2 and H_2S in the ionic liquid 1-ethyl-3-methylimidazolium tris (pentafluoroethyl) trifluorophosphate[J]. The Journal of Chemical Thermodynamics, 2013, 67: 55-62.

[3] Liu X, Huang Y, Zhao Y. Ionic liquid design and process simulation for decarbonization of shale gas[J]. Industrial & Engineering Chemistry Research, 2016, 55(20): 5931-5944.

[4] Wen G, Zhang X, Geng X. Liquid–liquid extraction of butanol from heptane+ butanol mixture by ionic liquids[J]. Journal of Chemical & Engineering Data, 2017, 62(12): 4273-4278.

[5] Xu X, Wen G, Ri Y. Liquid-liquid equilibrium measurements and correlation for phase behaviors of alcohols+ heptane+ ILs ternary systems[J]. The Journal of Chemical Thermodynamics, 2017, 106: 153-159.

[6] Wu X, Liu Z, Huang S. Molecular dynamics simulation of room-temperature ionic liquid mixture of [bmim][BF_4] and acetonitrile by a refined force field[J]. Physical Chemistry Chemical Physics, 2005, 7(14): 2771-2779.

[7] Liu Z, Huang S, Wang W. A refined force field for molecular simulation of imidazolium-based ionic liquids[J]. The Journal of Physical Chemistry B, 2004, 108(34): 12978-12989.

[8] Pan Q, Shang X, Li J. Energy-efficient separation process and control scheme for extractive distillation of ethanol-water using deep eutectic solvent[J]. Separation and Purification Technology, 2019, 219: 113-126.

[9] Mullins E, Oldland R, Liu Y A. Sigma-profile database for using COSMO-based thermodynamic methods[J]. Industrial & Engineering Chemistry Research, 2006, 45(12):4389-4415.

Chapter 3

Membrane Separation Process

3.1 Introduction

Membrane separation is a new separation technology that emerged in the early 20th century and rose rapidly after 1960s. Membrane separation technology has the function of separation, concentration, purification and has energy efficient and environmentally friendly features. Therefore, it has been widely used in medicine, biology, environmental protection, chemical, metallurgy, energy, petroleum, water treatment, electronics, bionics and other fields. It has produced enormous economic and social benefits and has become one of the most important methods in today's separation science. The membrane used in the process is a material with selective separation function. The process of separating, purifying and concentrating different components of a feed liquid by selective separation using membranes is regarded as membrane separation. It differs from conventional filtration in that the membrane can separate substance in molecular range, and the process is a physical process that does not require phase changes and additives.

3.2 Principle of Membrane Separation

Membrane separation technology is a new separation technique that uses a specially manufactured, selectively permeable membrane to separate, purify, and concentrate the mixture under external force. It is sieved according to the physical properties of the mixture. The rate of substances passing through the separation membrane (dissolution rate) depends on the difference in chemical properties among the membrane materials. The rate of diffusion is related to the molecular weight of the substance in addition to the chemical properties. The higher the velocity, the shorter the time required to permeate the membrane. The greater the velocity difference between components passing through the membrane in the mixture, the higher the separation efficiency.

3.3 Separation of DMSO-water Using Membrane

膜分离过程的模拟

(1) Starting Aspen Plus

① Click **Start** and then select Programs.

② Select **Aspen Tech | Process Modeling <version> | Aspen Plus | Aspen Plus <version>**.

③ Select **Chemical Processes | chemicals with Metric Units** or **Chemical Processes | Specialty Chemicals with Metrics Units**.

④ Select Blank Simulation then click **Create**.

For this customized simulation, Aspen Plus will handle everything but the calculations correspond to the ultrafiltration process itself.

(2) Specifying the Properties

① On the **Components | Specifications | Selection** sheet, type WATER and press **Enter** on the keyboard in the first cell under Component ID, as shown in Figure 3.1.

② In the next Component ID field, type DMSO and press **Enter** on the keyboard.

③ Go to the **Methods | Specifications | Global** sheet and select NRTL in the Base method field, as shown in Figure 3.2.

④ Go to **Methods | Parameters | Binary Interaction** sheet and select NRTL-1, as shown in Figure 3.3.

⑤ On the Home tab of the ribbon, in Units, make sure that the unit sets is METCBAR.

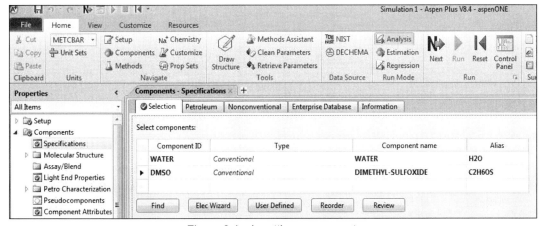

Figure 3.1   Inputting components

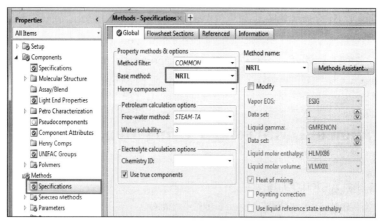

Figure 3.2 Physical property method

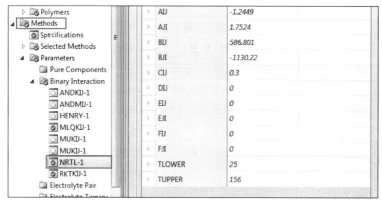

Figure 3.3 Determining interaction parameters

(3) Building the Process Flowsheet

① Click **Next**. Select Go to Simulation environment and click **OK** to go to the simulation environment, as shown in Figure 3.4.

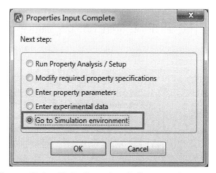

Figure 3.4 Entering simulation environment

② From the Model Palette, click **User Models** tab.

③ Click the arrow next to the User2 block icon to display all the User2 model icons. Then select FILTER, as shown in Figure 3.5.

④ Create one input stream and two product streams.

⑤ Name the input stream FEED by selecting the stream or its label, right click and select Rename Stream.

⑥ Name the first product stream you created RETENTAT. Name the second product stream you created PERMEATE.

⑦ Name the block MEMBRANE, as shown in Figure 3.6.

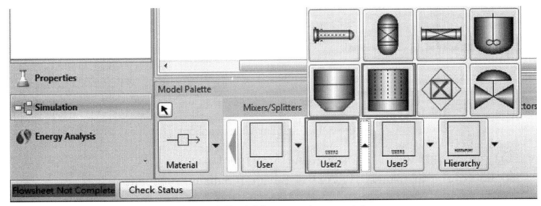

Figure 3.5   Adding a custom module

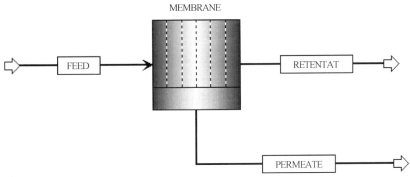

Figure 3.6   Connection process

(4) Entering Title, Components, Base Method, and Feed Specifications

① Go to the **Setup | Specifications | Global** sheet and input a title.

② Go to the **FEED | Input | Mixed** sheet, as shown in Figure 3.7.

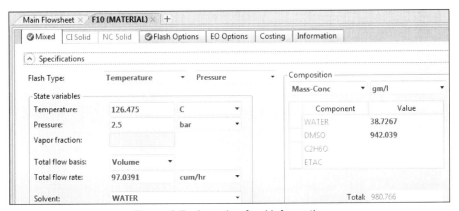

Figure 3.7   Inputting feed information

③ Enter the related parameters as shown in Figure 3.7.

④ Click the Flash Options tab to open the **FEED** | **Input** | **Flash Options** sheet. In the Valid phases field, select Liquid-Only, as shown in Figure 3.8.

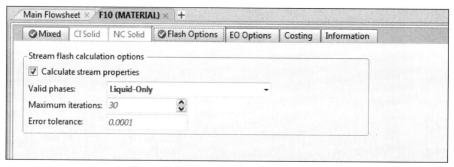

Figure 3.8 Determining flash options

(5) Entering Excel Path and User Array Data

① Go to the **MEMBRANE (User2)** | **Setup** | **Subroutines** sheet.

② In the Excel file name area, input the name of the future Excel file. For example, you might input MEMCALC1.XLS, as shown in Figure 3.9.

③ Click the User Arrays tab to open the **MEMBRANE (User2)** | **Setup** | **User Arrays** sheet, as shown in Figure 3.10.

④ In the Number of parameters area, input the related parameters as indicated below.

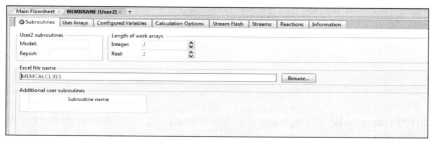

Figure 3.9 Entering the excel file name

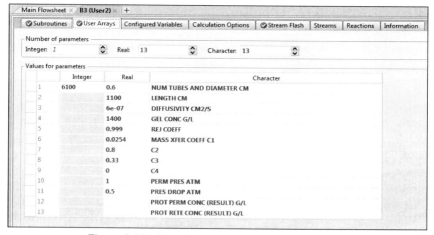

Figure 3.10 Inputting custom module parameters

(6) Setting up Product Stream Flash

① Click the **Stream Flash** tab to open the **MEMBRANE (User2)** | Setup | Stream Flash sheet, as shown in Figure 3.11.

② In the Stream field, select RETENTAT. In the Flash type field, select Temperature & pressure. In the Stream field, select PERMEATE. In the Flash type field, select Temperature & pressure, as shown in Figure 3.11. In a word, the Flash type of two streams RETENTAT and PERMEATE should be set as Temperature & pressure.

③ Save your Aspen Plus file as membrane.apw.

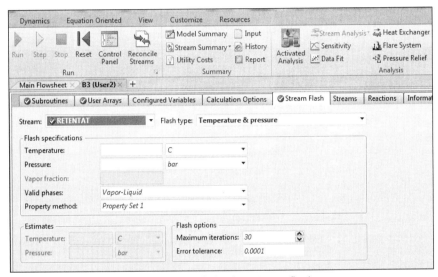

Figure 3.11 Selecting stream flash

(7) Copying and Examining the Excel Template

An Excel template is available to expedite the specification process. For Excel 97 and advanced version, use this template: Aspen Tech/Aspen Plus V10.0/Engine/User/userxlTemplate.xls.

① Place one of the templates in the folder you specified on the Aspen Plus **MEMBRANE (User2)** | Setup | Subroutines sheet.

② Open the template. Change the label, as shown in Figure 3.12. The four data sheets and the data will be filled after the simulation running.

Figure 3.12 Openning the template

(8) Editing Excel Sheets

① Edit the Aspen_ IntParams sheet, as shown in Figure 3.13. The NTUBES entry in cell C2 is strictly necessary. The entries in the first two columns are dummies,

Aspen Plus will fill in these cells when you run the simulation. Cell B2 will contain the integer parameter from the Aspen Plus User Arrays sheet.

	A	B	C	D	E	F	G	H
1	INTPARAM		1 DEFINED AS					
2		1	1 NTUBES					
3								
4								

Aspen_IntParams / Aspen_RealParams / Aspen_Output / Aspen_Input

Figure 3.13   Editing the Aspen_IntParams sheet

② Click **NTUBES** and select Name a range.
③ Edit variables to make them refer to cell B2.
④ Edit the Aspen_RealParams sheet, as shown in Figure 3.14.

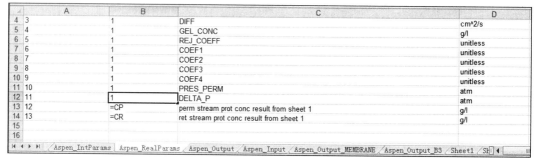

Figure 3.14   Editing the Aspen_RealParams sheet

⑤ Edit variables in the third column to make them refer to the second column except cell C13 and cell C14.
⑥ Edit the Aspen_Input sheet, as shown in Figure 3.15.

	A	B	C	D	E
1	INPUT	FEED STREAM	units	defined as	
2	WATER		1 kmol/s	WATER_FEED	
3	DMSO		1 kmol/s	PROT_FEED	
4	TOTFLOW		1 kmol/s	TOT_FEED	
5	TEMP		1 K	TEMP_FEED	
6	PRES		1 N/m^2	PRES_FEED	
7	ENTHALPY		1 J/kg		
8	VAP FRAC		1 molar		
9	LIQ FRAC		1 molar		
10	ENTROPY		1 J/kg-K		
11	DENSITY		1 kg/m^3	DENS_FEED	
12	MOLE WT		1 kg/kmol	MW_FEED	
13					
14					
15					

Aspen_IntParams / Aspen_RealParams / Aspen_Output / Aspen_Input

Figure 3.15   Editing the Aspen_Input sheet

⑦ Edit variables in the fourth column to make them refer to the second column.
⑧ Edit the Aspen_Output sheet, as shown in Figure 3.16. Click the icon in the top left corner. Then select **Excel Options** | **Advanced** and select the Show formulas in cells instead of their calculated results check box. The unit column

is optional.

	A	B	C	D
1	OUTPUT	RETENTAT	PERMEATE	units
2	WATER	=WATER_FEED -C2	=FP*RHO/3600/MWW	kmol/s
3	DMSO	=PROT_FEED - C3	=FP*CP/1000/3600/MWP	kmol/s
4	TOTFLOW	=TOT_FEED - C4	=C2+C3	kmol/s
5	TEMP	=TEMP_FEED	=TEMP_FEED	K
6	PRES	=PR	=PRES_PERM*101325	N/m^2
7	ENTHALPY	0	0	J/kg
8	VAP FRAC	0	0	molar
9	LIQ FRAC	0	0	molar
10	ENTROPY	0	0	J/kg-K
11	DENSITY	0	0	kg/m^3
12	MOLE WT	0	0	kg/kmol

Figure 3.16   Editing the Aspen_Output sheet

⑨ Edit Sheet1, as shown in Figure 3.17. Edit variables in the second column to make them refer to the first column.

	A	B	C	D
1	Model Input			
2	MWP	78.13	kg/kmol prot mole wgt	
3	MWW	18.01528	kg/kmol water mole wgt	
4	MU	0.00468977	g/cm/s viscosity	
5	PMF	=PROT_FEED*2:2	prot mass flow feed stream	
6	TMF	=TOT_FEED*MW_FEED	total mass flow feed stream	
7	Feed Stream			
8	RHO	=DENS_FEED/1000	g/cm3 density	
9	FIN	=TOT_FEED*MW_FEED/8:8*3600	l/hr feed stream flow rate	
10	CIN	=5:5/6:6*8:8*1000	g/l feed stream prot conc	
11	PIN	=PRES_FEED/101325	atm feed stream pressure	
12	Intermediate			
13	UAVE	=9:9/(PI()/4*DIAM^2*NTUBES)*1000/3600	cm/s bulk average velocity	
14	RE	=DIAM*13:13*8:8/4:4	reynolds number (unitless)	
15	SC	=4:4/8:8/DIFF	schmidt number (unitless)	
16	MT_1	=COEF1	Mult Term 1	
17	MT_2	=14:14^COEF2	Mult Term 2	
18	MT_3	=15:15^COEF3	Mult Term 3	
19	MT_4	=(DIAM/LEN)^COEF4	Mult Term 4	
20	Results			
21	CP	=GEL_CONC*(1-REJ_COEFF)	g/l perm stream prot conc	
22	K	=16:16*17:17*18:18*19:19*DIFF/DIAM/100	m/s mass xfer coefficient	
23	J	=22:22*LN((GEL_CONC-21:21)/(10:10-21:21)	m/s volumetric flux	
24	FP	=23:23*DIAM*LEN*PI()*NTUBES*100*3600/	l/hr perm stream flow rate
25	PR	=(11:11-DELTA_P)*101325	N/m^2 ret stream pressure	
26	CR	=(10:10-24:24/9:9*21:21)/(1-24:24/9:9)	g/l ret stream prot conc	

Figure 3.17   Editing Sheet1

⑩ Click the icon in the top left corner. Then select **Excel Options | Advanced** and select the Show formulas in cells instead of their calculated results check box. The unit column is optional.

⑪ Save as me12mcalc1.xls. This file will be altered by Aspen Plus.

⑫ Run the simulation.

⑬ View simulation results, as shown in Figure 3.18.

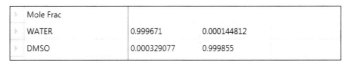

Figure 3.18   Simulation results

Chapter 3   Membrane Separation Process

Exercises

1. Membrane separation module is used to separate water and glycerol. Feed flow rate is 100 kmol/h with the composition of 73% (mole fraction) glycerol and 27% water, temperature is 145.4℃, pressure is 1 atm. The purity of two products is 99.9%. Conduct the simulation. (The case is from Journal of the Taiwan Institute of Chemical Engineers, 2018, 91: 251-265.)

2. Membrane separation module is used to separate water and ethanol. Feed flow rate is 23500 kg/h with the composition of 62.2% (mass fraction) ethanol and 37.8% water, temperature is 125.2℃, pressure is 358 kPa. The purity of two products is 99.9% (mole fraction). Conduct the simulation. (The case is from Journal of Chemical Technology & Biotechnology, 2019, 94(4): 1041-1056.)

References

[1] Mericq J P, Laborie S, Cabassud C. Evaluation of systems coupling vacuum membrane distillation and solar energy for seawater desalination[J]. Chemical Engineering Journal, 2011, 166(2): 596-606.

[2] Legay M, Allibert Y, Gondrexon N, et al. Experimental investigations of fouling reduction in an ultrasonically-assisted heat exchanger[J]. Experimental Thermal and Fluid Science, 2013, 46: 111-119.

Chapter 4

Heat-integration and Thermally Coupled Distillation

4.1 Introduction

In chemical industry, distillation operation is an important step, but it is also a high energy consumption process, the energy consumption of the whole distillation process accounts for a large part of the total energy consumption.

The heat-integrated distillation is one of the methods to reduce energy consumption in distillation process. In the heat-integrated distillation process, heat transfer is carried out between the flow at the top of one distillation column with higher temperature and the flow at the bottom of another distillation column with lower temperature.

In pressure-swing distillation, distillation columns at different pressures produce different temperatures. Therefore, the use of heat-integration is often economically attractive in pressure-swing systems.

The heat-integration includes the full heat-integration and the partial heat-integration. In the full heat-integration, there is only one steam heat reboiler in high pressure column (HPC) and one water cooling condenser in low pressure column (LPC). In addition to these facilities, an extra heat exchanger could be designed as a condenser in HPC or a reboiler in LPC. Thus, the heat removal in the condenser of the HPC is exactly equal to the heat input in the reboiler of the LPC.

Alternatively, in the partial heat-integration, the heat removal in the condenser of the HPC is not equal to the heat input in the reboiler of the LPC. An auxiliary reboiler or condenser is required. There is an economic and controllable trade-off in the choice of methods. The full heat-integration is usually more economical in the facility cost and energy cost. The partial heat-integration usually results in a better dynamic control because there are more control degrees of freedom.

Because the partial heat-integration has one more auxiliary condenser or reboiler than the full heat-integration, and the optimization of the partial heat-integration process involves multiple optimization variables. Therefore, the optimization of the full heat-integration is different from that of the partial heat-integration. For example, in the process of the full heat-integration, the reflux ratio in T1 usually needs to be changed to meet the global design specifications that the heat of the LPC kettle is equal to the top heat of the HPC. Therefore, the degrees of freedom of optimization variables in the full heat-integration process are less than those in the partial heat-integration process and the reflux ratio in T1 no longer needs to be optimized in the full heat-integration, and the optimization process of other variables are the same as those of the partial heat-integration.

4.2 Steady-state Simulation of THF-methanol System with Heat-integration

The simulation of the THF-methanol system will be carried out first. A comparison of systems without heat-integration, with the partial heat-integration and with the full heat-integration will be presented.

4.2.1 Simulation without Heat-integration

The flowsheet without heat-integration is shown in Figure 4.1.

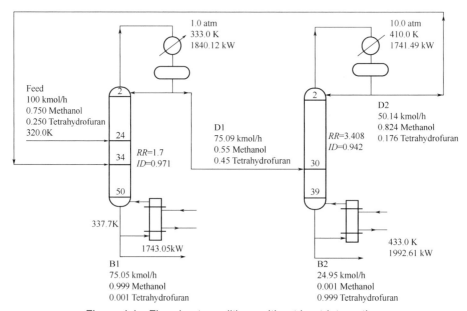

Figure 4.1 Flowsheet conditions without heat-integration

(1) Input the components, as shown in Figure 4.2.

Figure 4.2 Inputting the components

(2) Select NRTL as physical property method and view the binary interaction parameters, as shown in Figure 4.3.

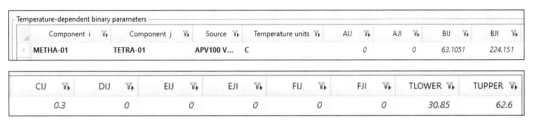

Figure 4.3 Viewing the binary interaction parameters

(3) Input the calculated parameters and create simulation process without loops, as shown in Figure 4.4.

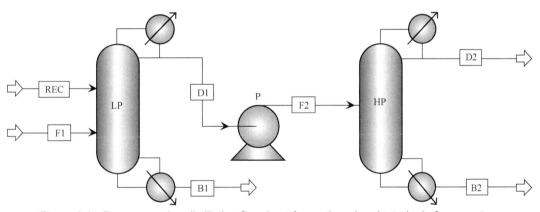

Figure 4.4 Pressure-swing distillation flowsheet for methanol and tetrahydrofuran system

(4) Run and view the results of each stream, as shown in Figure 4.5.

(5) Add design specifications. The type of design specification is mole purity. The target is 0.999, as shown in Figure 4.6.

(6) Run and view the results of each stream, as shown in Figure 4.7.

(7) Input the information of stream D2 into stream REC, reset and run the simulation, as shown in Figure 4.8. View the results of each stream in Figure 4.9.

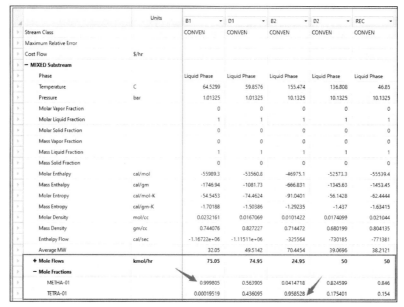

Figure 4.5　Simulation results of each stream

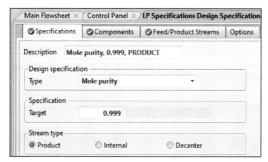

Figure 4.6　Adding design specifications of the high-pressure column

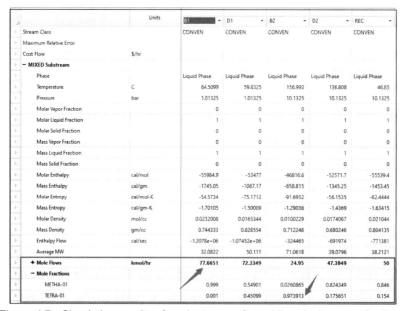

Figure 4.7　Simulation results of each stream after adding design specifications

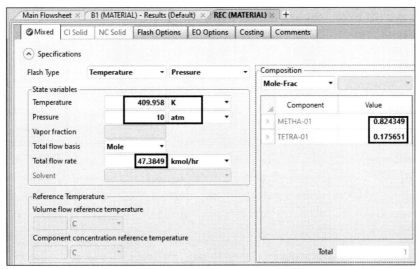

Figure 4.8　Inputting the information of stream REC

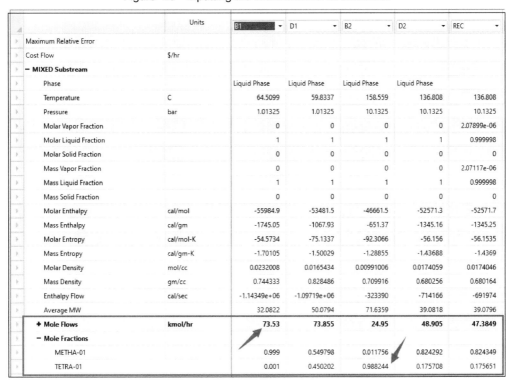

Figure 4.9　Simulation results after inputting the information of stream REC

(8) Repeat step (7) until the separation requirement is reached, as shown in Figure 4.10.

(9) Remove the design specification and merge D2 with REC. Then reset and run the simulation, as shown in Figure 4.11. View the results of each stream in Figure 4.12.

(10) At last, check the heat duties of condenser in HPC and reboiler in LPC, as shown in Figure 4.13.

	Units	B1	D1	B2	D2	REC
Stream Class		CONVEN	CONVEN	CONVEN	CONVEN	CONVEN
Maximum Relative Error						
Cost Flow	$/hr					
− MIXED Substream						
Phase		Liquid Phase	Liquid Phase	Liquid Phase	Liquid Phase	Liquid Phase
Temperature	C	64.5099	59.8347	159.846	136.809	136.809
Pressure	bar	1.01325	1.01325	10.1325	10.1325	10.1325
Molar Vapor Fraction		0	0	0	0	0
Molar Liquid Fraction		1	1	1	1	1
Molar Solid Fraction		0	0	0	0	0
Mass Vapor Fraction		0	0	0	0	0
Mass Liquid Fraction		1	1	1	1	1
Mass Solid Fraction		0	0	0	0	0
Molar Enthalpy	cal/mol	-55984.9	-53485	-46539.6	-52568.2	-52568.2
Mass Enthalpy	cal/gm	-1745.05	-1068.55	-645.784	-1344.43	-1344.42
Molar Entropy	cal/mol-K	-54.5734	-75.1034	-92.78	-56.1765	-56.1765
Mass Entropy	cal/gm-K	-1.70105	-1.50045	-1.28742	-1.43671	-1.43671
Molar Density	mol/cc	0.0232008	0.0165508	0.0098239	0.0173998	0.0173998
Mass Density	gm/cc	0.744333	0.82843	0.707978	0.680346	0.680346
Enthalpy Flow	cal/sec	-1.16714e+06	-1.11566e+06	-322545	-732207	-732213
Average MW		32.0822	50.0539	72.0668	39.1009	39.1009
+ Mole Flows	kmol/hr	75.0504	75.0933	24.95	50.1433	50.1437
− Mole Fractions						
METHA-01		0.999	0.550434	0.00100027	0.823817	0.823816
TETRA-01		0.001	0.449566	0.999	0.176183	0.176184

Figure 4.10 The stream information of D2 and REC

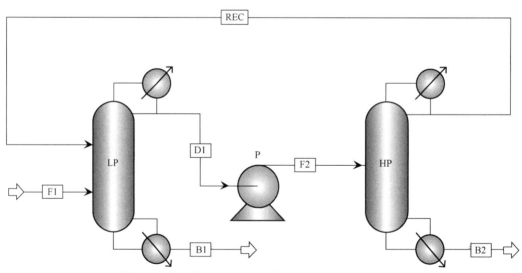

Figure 4.11 The flowsheet after merging D2 with REC

4.2.2 Simulation with Partial Heat-integration

The paper published by Wang et al discussed the pressure-swing distillation of THF-methanol system using the heat-integration. The optimum design of a heat-integrated two column system was determined by minimizing the total energy

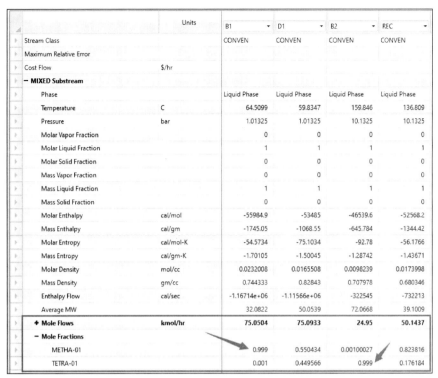

Figure 4.12 Final simulation results of each stream

Condenser / Top stage performance		
Name	Value	Units
Temperature	136.809	C
Subcooled temperature		
Heat duty	-1741.43	kW
Subcooled duty		
Distillate rate	50.1433	kmol/hr
Reflux rate	170.889	kmol/hr
Reflux ratio	3.408	

(a)

Reboiler / Bottom stage performance		
Name	Value	Units
Temperature	64.5099	C
Heat duty	1743.01	kW
Bottoms rate	75.0504	kmol/hr
Boilup rate	177.939	kmol/hr
Boilup ratio	2.37093	

(b)

Figure 4.13 The heat duties of condenser in HPC and reboiler in LPC

consumption (the sum of the two reboilers duties). They used the heat removed from the condenser of the HPC as a portion of the heat input to the base of the

LPC. The remaining heat was provided by an auxiliary reboiler, which was heated by low pressure steam. The reboiler of the HPC was driven by high-pressure steam. Thus, a partially heat integrated system was studied. An auxiliary reboiler provided an additional control degree of freedom, so both tray temperature and reflux ratio could be independently manipulated in each column.

The flowsheet is shown in Figure 4.14. The base case has 100 kmol/h feed with a composition of 25% (mole fraction) THF. Operating pressures in the two columns are 1 atm and 10 atm. The number of stages in the two columns are 51 and 40 with the condenser as stage1. The methanol product stream B1 from the bottom of the LPC contains 0.1% (mole fraction) THF and the product stream B2 from the bottom of the HPC contains 0.1% (mole fraction) methanol.

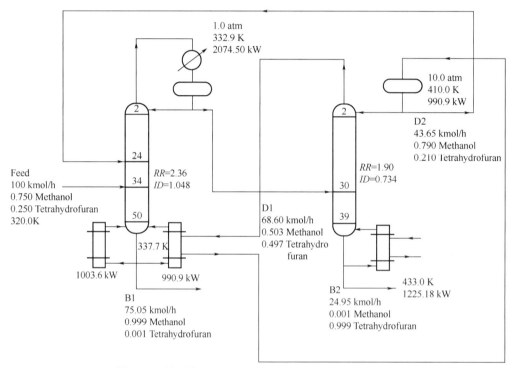

Figure 4.14 Flowsheet with the partial heat-integration

The Aspen design spec and vary capability are used to hold the two product purity by manipulating the flow rates of the two bottom streams. The reflux ratios in the two columns are the design optimization variables. They vary until the minimum reboiler energy consumption is determined (the sum of H1 and H2). The optimum values of the reflux ratios are $RR1=2.36$ and $RR2=1.9$. The corresponding total energy consumption is 3219.68 kW ($H1$=1994.5 kW and $H2$=1225.18 kW).

Under these conditions, the heat removal in the condenser of the HPC is 990.9 kW. This suggests that a partially heat-integrated flowsheet could be used. A condenser/reboiler could transfer 990.9 kW energy into the LPC. The temperature difference between the condensing vapor from the HPC (410.0 K) and the boiling liquid at the bottom of the LPC (337.7 K) is adequate for heat integration with a reasonable area heating operating condition. The remaining energy requirement in the low-pressure column can be satisfied by a steam-heated auxiliary reboiler supplying 1994.5-990.9=1003.6 kW.

4.2.3 Simulation with Full Heat-integration

Figure 4.15 gives the flowsheet of full heat-integration with feed flowrate of 100 kmol/h and a feed composition of 25 % (mole fraction) THF. The total number of stages in each column is the same as the flowsheet of the partial heat-integration. At this point, the heat duty of the condenser/reboiler is known. All the temperatures throughout both columns are also known.

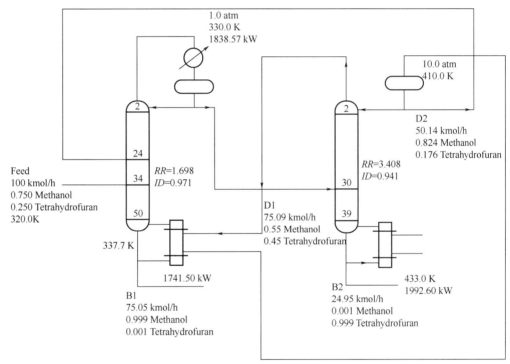

Figure 4.15　Flowsheet with the full heat-integration

In this case, auxiliary reboiler and condenser are not used. The energy input is only the heat-transfer duty in the HPC E2. This "neat" configuration minimizes the total energy consumed. However, a control degree of freedom is lost since the heat inputs to the LPC H1 cannot be independently set, it must be equal to the heat removal rate of E2.

The design spec and vary feature in each of the column blocks in Aspen Plus V10 are used to adjust the bottom flowrate to achieve the desired product purity in each column. The specifications for the product purity are the same in all cases. The methanol product from the base of the LPC B1 has a composition of 0.1%

(mole fraction) THF. The THF product from the base of the HPC B2 has a composition of 0.1% (mole fraction) methanol. With the full heat-integration, the heat removal in the condenser of the HPC E2 must be equal to the heat input to the reboiler of the LPC H1.

Achieving steady-state simulations in Aspen Plus that rigorously capture the neat heat-integration requires the use of a "Flowsheet design spec" to make $H1=E2$. Figure 4.16~Figure 4.20 show how this is set up. Reflux ratio in the LPC $RR1$ is selected as the manipulated variable (Figure 4.19). Thus, the reflux ratio of the HPC is varied by the flowsheet design spec to make $H1=E2$ (Figure 4.20). With the reflux ratio fixed in the LPC, there are three variables ($B1$, $B2$, and $RR2$) being used to drive the two product compositions to their desiring specifications and make the heat duties equal in magnitude but opposite in sign. The next step of the design is to vary the reflux ratio of the LPC to find the value that minimizes the heat input to the reboiler of the HPC E2. Specific steps are as follows.

(1) Find the option Flowsheeting Options. Choose the Design Specs option and set up a new design spec using the default name of system DS-1, as shown in Figure 4.16.

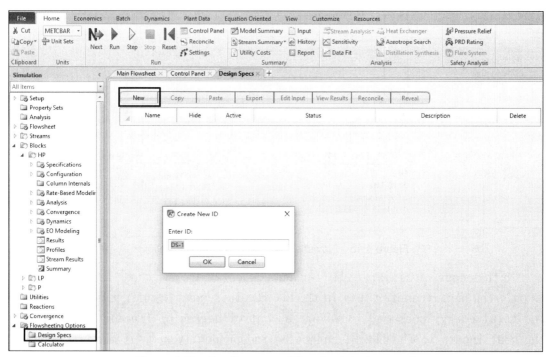

Figure 4.16   Creating a new "Flowsheet design spec"

(2) Next, define the variables. Add three variables using the names of H1, E2 and Q. The definition of these three variables are shown in Figure 4.17.

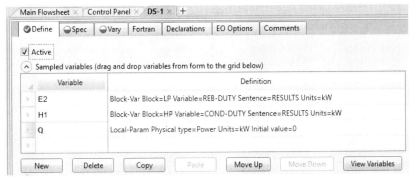

Figure 4.17 Defining the variables

(3) Input the parameters of Q. The target is 0 and the tolerance is 0.00001, as shown in Figure 4.18.

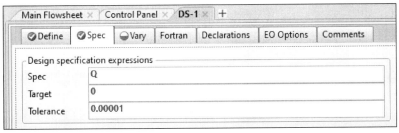

Figure 4.18 The parameters of Q

(4) Carry out the point which makes $H1=E2$ by controlling the manipulated variable and choose the MOLE-RR of the LPC as the manipulated variable, as shown in Figure 4.19.

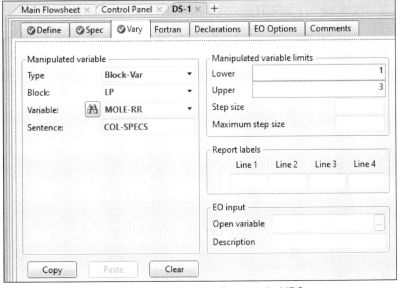

Figure 4.19 Varying reflux ratio in HPC

(5) In the Fortran option, input the executable Fortran statement $Q=H1+E2$, as shown in Figure 4.20.

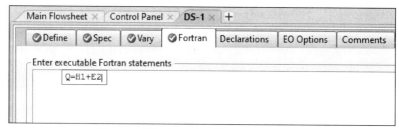

Figure 4.20 Inputting Fortran statement

(6) Run the simulation and view the results in Figure 4.21.

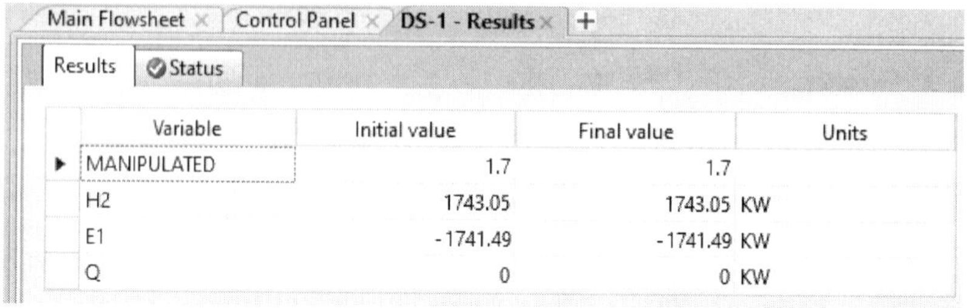

Figure 4.21 The results of design spec

The comparison of three flowsheets:

If the two columns have independent reboilers and condensers, both temperature and reflux ratio can be manipulated in each column. In addition, the pressures in both columns can be controlled. Therefore, the pressure in the high-pressure column does not float with operating conditions, which is advantageous for temperature control. Comparing the flowsheet in Figure 4.1 and Figure 4.14, the columns are exactly the same, but the LPC column in the flowsheet of Figure 4.1 has only one reboiler. The total energy consumption in this design (3735.66 kW) is the sum of H1 (1743.05 kW) and H2 (1992.61 kW). This should be compared with the energy consumption of the full heat-integration system (1992.61 kW) and the energy consumption of the partial heat-integration system (2228.78 kW).

These results clearly demonstrate that heat-integration can result in significant reductions in energy consumption. However, as we demonstrate in the next section, the dynamic controllability of the fully heat-integrated system is not as good as that of the no heat-integrated system.

4.3 Thermally Coupled Distillation Process

As is known to all, distillation process is an important operating unit of the chemical process, and it is a typical representative of high energy consumption and low energy efficiency. Most energy is consumed in the distillation unit in the whole petrochemical process, while the highest energy utilization rate of distillation can only be about 15%, a large amount of energy is wasted, which seriously restricts the sustainable development of China's economy. However, at the same time, distillation is the most widely used operating unit in chemical industry, therefore, the reform and innovation of distillation technology has received extensive attention. In recent years, it has been the forefront and difficulty of research in the field of chemical industry and process control.

The low energy efficiency of distillation process is mainly determined by its complex process characteristics. The research of distillation energy-saving mainly concentrates in two aspects, the first is the improvement of process flow and design in the distillation process. By exploring the thermal energy loss of the entire process, a more reasonable process flow is designed, which improves the heat reuse of the entire process. So as to achieve the purpose of reducing thermal energy loss. The second is the optimization of process operation. On the one hand, it can optimize the current operating point and make the current system in a higher running under the condition of energy efficiency. On the other hand, it can improve the performance of control system and make the whole system run smoothly for a long time as far as possible, it can also reduce the energy loss of the whole system effectively.

In conventional distillation process, a large amount of energy is invested in the reboiler, while it is wasted in the condenser, which is the main reason for the low energy efficiency of the whole distillation process. The internal thermally coupled distillation technology also starts from this main reason to seek for the greater efficiency of heat utilization. Through thermally coupling between rectifying section and stripping section and a proper operation optimization, the load of condenser and reboiler can be greatly reduced, even zero load of condenser and reboiler in normal operation. The energy efficiency can be greatly improved because of no need of condenser and reboiler comparing with conventional distillation process.

Heat pump distillation, as one of the thermally coupled distillation technologies, is an effective energy saving technology for separating near boiling point substances, which has been widely used in practical industrial production. The basic principle of this process is to recompress the steam at the top of the column. The temperature and pressure of the steam are rising to heat the bottom of the column, which can reduce the load of reboiler. This process is suitable for the separation process with small temperature difference at the top and bottom of the column, small compression and small compression power.

In addition, the pressure-swing thermally coupled distillation technology is widely used in high energy consumption of the separation process, which the boiling point is close. Its basic principle is that the conventional rectification column is divided into two different pressure column section. HPC is the distillation section, LPC is the stripping section. Using the high temperature steam of HPC to heat the bottom liquid of LPC, the thermally coupled operation can greatly save energy.

The thermally coupled distillation can save more energy than conventional distillation, which has great application prospect. Due to the complex nonlinear dynamic characteristics of internal thermally coupled distillation process, such as strong coupling, strong ill-condition, strong asymmetry, and strong inverse response, its control design has been the bottleneck problem hindering the commercialization of the high efficiency and energy saving technology.

4.4 Energy-saving Thermally Coupled Ternary Extractive Distillation Process

The major intrinsic obstacle of extractive distillation is the high energy consumption. The paper published by Zhao et al discussed the ternary extractive distillation using thermally coupled distillation process. Two thermally coupled ternary extractive distillation processes were studied to separate the ternary azeotropic mixture tetrahydrofuran/ethanol/water using dimethyl sulfoxide (DMSO) as entrainer.

This section begins with the thermally coupled ternary extractive distillation process, and then presented specific operation.

The flowsheet is shown in Figure 4.22. The base case has 100 kmol/h feed with a composition of 30% (mole fraction) THF, 30% (mole fraction) ethanol and 40% (mole fraction) water. The operating pressure of the three columns is 0.7 atm, 0.4 atm and 0.15 atm respectively. And the numbers of stages in the columns are 60, 55 and 20 with the condenser as stage1. Product purity is 0.999 (mole fraction) THF in the stream D1 from the top of the T1 and 0.999 (mole fraction) ethanol in the stream D2 from the top of the T2, 0.999 (mole fraction) water in the stream D3 from the top of the T3 and 0.99999 (mole fraction) DMSO in the stream B3 from the bottom of T3.

The amount of DMSO in T1 and T2 are 25kmol/h and 35kmol/h. The reflux ratios of T1 and T3 are 2 and 1. In this example, the side production flow rate of the thermal coupling part of T1 is 80 kmol/h (gas phase). The separation of the mixture is achieved by using the heat and state of the stream itself, so as to reduce the reboiler of T2. The detailed operation is as follows.

(1) First, input the components as follows, TETRAHYDROFURAN, ETHANOL, WATER and DIMETHYL SULFOXIDE, as shown in Figure 4.23.

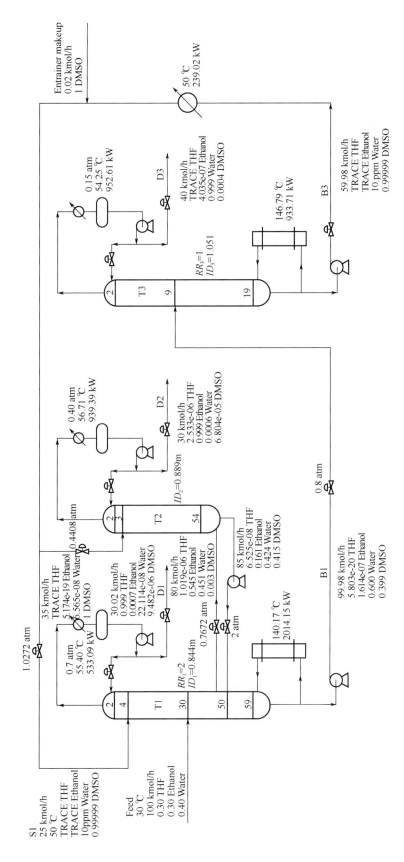

Figure 4.22 Flowsheet of thermally coupled extractive distillation

Chapter 4 Heat-integration and Thermally Coupled Distillation

Figure 4.23 Inputting the components

(2) The property method is selected as NRTL. The binary interaction parameters of these four materials are shown in Figure 4.24.

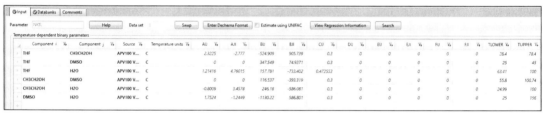

Figure 4.24 Viewing the binary interaction parameters

(3) Input the parameters of column T1, the flowsheet is shown in Figure 4.25.

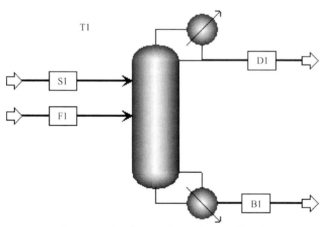

Figure 4.25 Separation column of THF

(4) Run and view the results in Figure 4.26.
(5) The thermally coupled columns are established in Figure 4.27.
(6) Run and view the results in Figure 4.28.
(7) Then, input the parameters of column T3 (used to separate water), as shown in Figure 4.29, and the results are shown in Figure 4.30.

	Units	D1	B1
− MIXED Substream			
Phase		Liquid Phase	Liquid Phase
Temperature	C	55.3908	95.0122
Pressure	bar	0.709275	1.11579
Molar Vapor Fraction		0	0
Molar Liquid Fraction		1	1
Molar Solid Fraction		0	0
Mass Vapor Fraction		0	0
Mass Liquid Fraction		1	1
Mass Solid Fraction		0	0
Molar Enthalpy	cal/mol	-50682	-60932.9
Mass Enthalpy	cal/gm	-703.039	-1427.18
Molar Entropy	cal/mol-K	-103.223	-57.4823
Mass Entropy	cal/gm-K	-1.43187	-1.34636
Molar Density	mol/cc	0.0117276	0.0204167
Mass Density	gm/cc	0.845439	0.871681
Enthalpy Flow	cal/sec	-422632	-1.60761e+06
Average MW		72.0899	42.6945
+ Mole Flows	kmol/hr	30.02	94.98
− Mole Fractions			
THF		0.999334	7.60234e-10
CH3CH2OH		0.000656738	0.315648
DMSO		9.48227e-06	0.26321
H2O		4.58827e-09	0.421141

Figure 4.26 The simulation results of THF

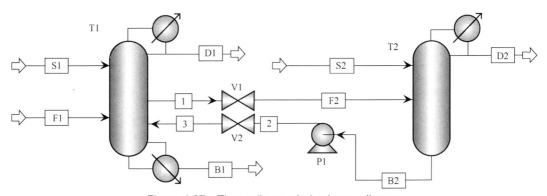

Figure 4.27 Thermally coupled columns diagram

(8) Next, add the entrainer, calculate circulation quantity and adjust the pressure, as shown in Figure 4.31 and view results in Figure 4.32.

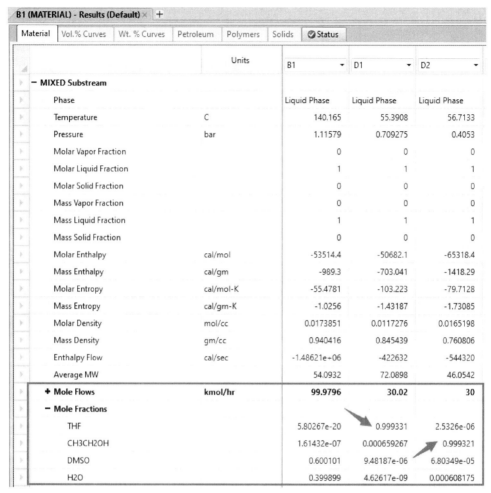

Figure 4.28 Simulation results of each stream

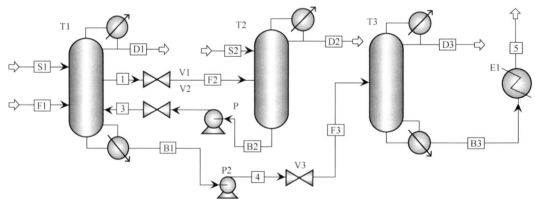

Figure 4.29 The process of solvent recovery

(9) At last, the process is linked by the circular streams in Figure 4.33 and the results of each stream are shown in Figure 4.34.

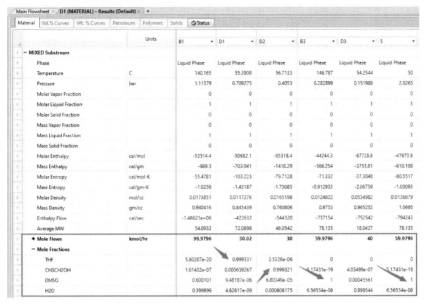

Figure 4.30　The results of each stream

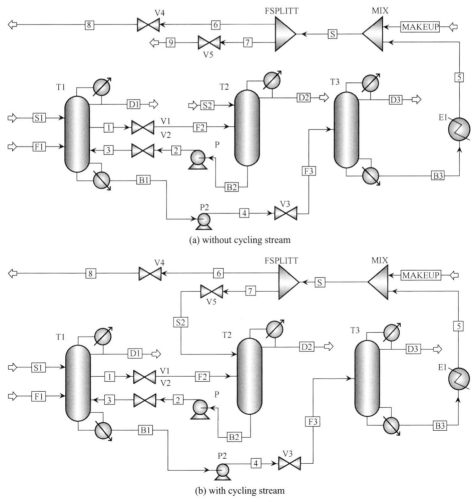

(a) without cycling stream

(b) with cycling stream

Figure 4.31　The process of solvent recovery

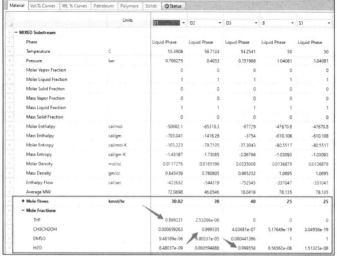

Figure 4.32　Simulation results of each stream

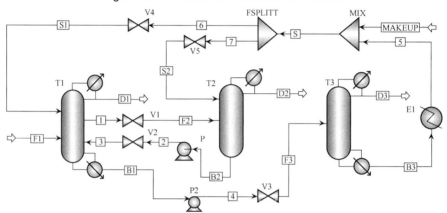

Figure 4.33　Linking the circular streams

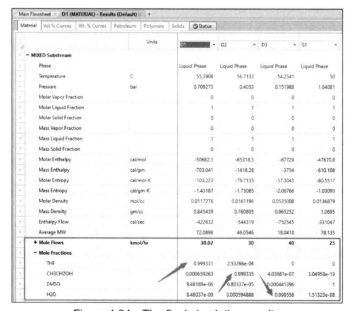

Figure 4.34　The final simulation results

Exercises

1. The separation process of *n*-heptane and isobutanol via pressure-swing distillation was designed. Detailed information about the steady-state process is shown in Figure 4.35. Complete the partial heat-integration. (The case is from Zhang. Process optimization and control strategy of pressure-swing distillation for separating an azeotropic mixture of *n*-heptane and isobutanol [D]. Qingdao: Qingdao University of Science and Technology, 2016.)

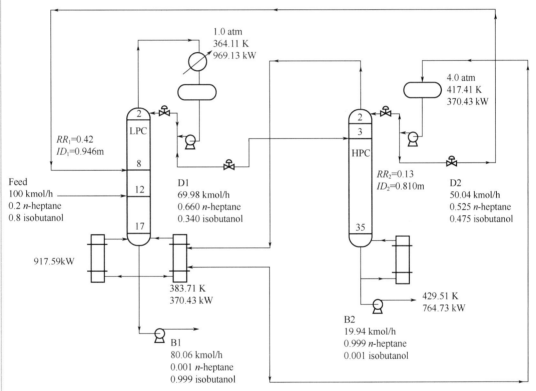

Figure 4.35 Flowsheet conditions with the partial heat-integration

2. The separation process of *n*-heptane and isobutanol via pressure-swing distillation was designed. Detailed information about the steady-state process is shown in Figure 4.36. Complete the full heat-integration. (The case is from Zhang. Process optimization and control strategy of pressure-swing distillation for separating an azeotropic mixture of *n*-heptane and isobutanol [D]. Qingdao: Qingdao University of Science and Technology, 2016.)

3. The separation process of THF, water and ethanol via thermally coupled ternary extractive distillation was designed. Detailed information about the steady-state process is shown in Figure 4.37. Complete the thermally coupled process of the C2 and C3. (The case is from Energy, 2018, 148: 296-308.)

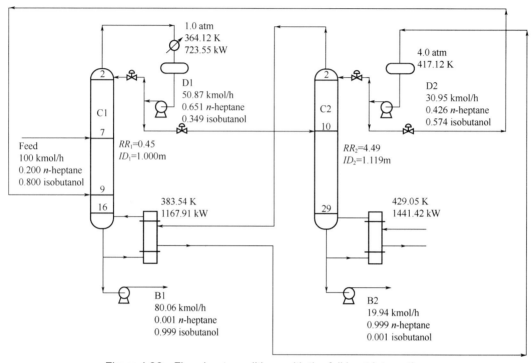

Figure 4.36 Flowsheet conditions with the full heat-integration

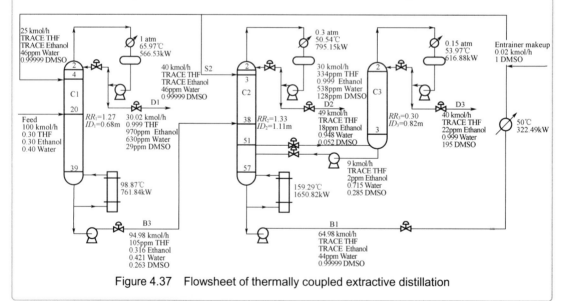

Figure 4.37 Flowsheet of thermally coupled extractive distillation

References

[1] Wang Y, Cui P, Zhang Z. Heat-integrated pressure-swing-distillation process for separation of

tetrahydrofuran/methanol with different feed compositions[J]. Industrial & Engineering Chemistry Research, 2014, 53(17): 7186-7194.

[2] Zhao Y, Ma K, Bai W, et al. Energy-saving thermally coupled ternary extractive distillation process by combining with mixed entrainer for separating ternary mixture containing bioethanol[J]. Energy, 2018, 148: 296-308.

[3] Lee J, Cho J, Kim D M, et al. Separation of tetrahydrofuran and water using pressure swing distillation: Modeling and optimization[J]. Korean Journal of Chemical Engineering, 2011, 28(2): 591-596.

[4] Kim Y H. Energy saving of benzene separation process for environmentally friendly gasoline using an extended DWC (divided wall column)[J]. Energy, 2016, 100: 58-65.

[5] Timoshenko A V, Anokhina E A, Morgunov A V, et al. Application of the partially thermally coupled distillation flowsheets for the extractive distillation of ternary azeotropic mixtures[J]. Chemical Engineering Research and Design, 2015, 104: 139-155.

Chapter 5

Heat Pump Distillation for Close-boiling Mixture

5.1 Introduction

With the development of the industrial revolution in the 19th century, people became interested in whether heat could be pumped from lower-temperature substances to higher-temperature substances. In the 1920s, Sadi Karnot, a French scientist, put forward the famous theory of Kano cycle, which became the beginning of heat pump technology. Thirty years later, British scientist L Kelvin put forward the concept that a cooling device can be used for heating, i.e. a "heat multiplier". He described the concept of heat pump first. After that, a large number of scholars and scientists began to study heat pump technology.

The heat pump can be considered as the reverse process of heat engine. The heat engine passes heat from the high temperature heat source to the low temperature heat source and it is used for external work. The heat pump needs to provide external power or drive energy to remove the heat from the low temperature substances and increase its temperature.

5.2 Main Forms of Heat Pump Distillation

Distillation unit operation is often used to separate homogeneous liquid mixtures, which is mainly based on the difference of relative volatility between homogeneous mixtures. In the process of distillation, some liquids in the distillation column are continuously vaporized and some gases are continuously condensed, so that the purity of products at the top and bottom of the column is qualified. In order to achieve vaporization and condensation, in the traditional distillation process, a condenser is installed at the top of the distillation column to remove part of the heat from the steam at the top of the column, so that the column can be refluxed; a reboiler is installed in the column to generate rising steam. The basic structure and energy transfer of the traditional distillation process are shown in Figure 5.1 (a).

In the operation of traditional distillation column, the steam condensation at the top of the column and the heating of the kettle need to be completed by utilities. It can be seen that distillation process is a process that requires both heating and refrigeration. In connection with the working principle of heat pump, it is very effective to realize energy saving by combining heat pump technology for distillation process operated in a single column. The basic structure and energy transfer of heat pump distillation process are shown in Figure 5.1 (b).

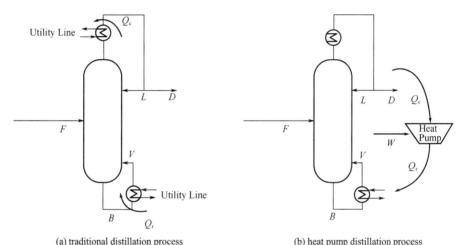

(a) traditional distillation process   (b) heat pump distillation process

Figure 5.1 The schematic diagram of distillation process and energy transfer

When the temperature difference between the top of the column and the bottom of the column is not large enough, the following three forms of heat pump rectification can be considered to achieve heat-integration.

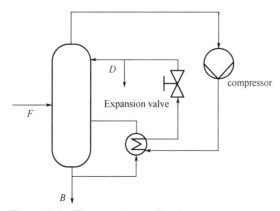

Figure 5.2 The top steam direct compression type

As shown in Figure 5.2, the steam at the top of the column is heated by the compressor and it is condensed and released in the reboiler. Then, it returns to the top of the column as reflux.

The liquid in the column is cooled and reduced in the condenser and evaporated into gas. After being pressurized by the compressor, the liquid returns to the bottom of the column as boiling steam, as shown in Figure 5.3.

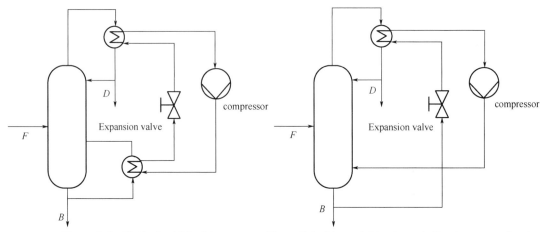

Figure 5.3 Kettle liquid flashing   Figure 5.4 The outside steam indirect compression type

The external fluid absorbs and evaporates into gas in the condenser. Then, the gas condenses and releases heat in the reboiler after being heated by the compressor, as shown in Figure 5.4.

5.3 Heat Pump Distillation Process of Binary System Close-boiling Mixture

Close-boiling binary system, as its name implies, is a mixture of two substances with similar boiling point. The relative volatility of this kind of system is generally close to 1 at atmospheric pressure. It takes a lot of energy to separate this kind of mixture by traditional single column distillation. In order to improve the energy utilization efficiency, many

热泵精馏实例

energy-saving technologies have been developed, including heat pump distillation, thermally coupled distillation, intermediate reboiler and intermediate condenser.

In addition, special distillation methods such as pressure swing distillation, extractive distillation column can be used to separate such mixtures. A large number of studies have shown that heat pump distillation has great potential for the separation of binary mixtures with near boiling point.

Example: the use of overhead steam direct compression heat pump distillation column to separate *n*-butanol-isobutanol is required. The purity of the top and bottom products is not less than 99% by adjusting the operating parameters of the distillation column (mass fraction), its specific operating parameters are as follows:

Feed pressure is 0.5 MPa, feed flow rate is 5000 kg/h, where the mass fraction of *n*-butanol is 45%, the mass fraction of isobutanol is 55%, the number of distillation column trays is 61, the feed stage is 34, the reflux ratio is 7.12, the bottom product flow rate is 276.84 kmol/h and the physical property method is WILSON.

The heat pump distillation column process is shown in Figure 5.5.

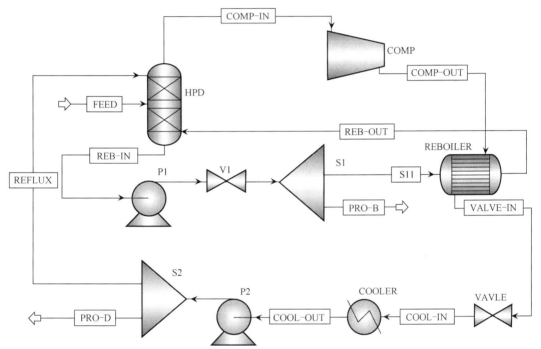

Figure 5.5 The optimal flowsheet of the MVRHP distillation for separating close-boiling mixture of *n*-butanol and isobutanol

(1) To facilitate process convergence and determine cell module parameters, a non-circulating heat pump rectification column process shown in Figure 5.6 is established, where the heat pump rectification column HPD uses the Columns| Rad Frac| PACKABS icon in the module palette.

The reflux of the top of the rectification column and the reflux of the reboiler are taken as the initial values of the stream REFLUX and the stream REB-OUT, respectively.

(2) Go to the Blocks| COLUMN| Results| Summary page, Blocks| COLUMN| Profiles| TPFQ page, Blocks| COLUMN| Profiles| Compositions page of the

conventional distillation column to view the temperature, pressure, flow rate and composition of the reflux and return stream, as shown in Figure 5.7 and Figure 5.8.

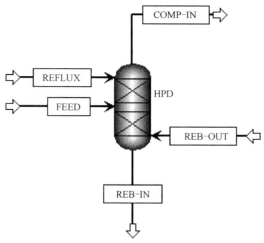

Figure 5.6 Establishing a non-circulating heat pump distillation column process

(a)

(b)

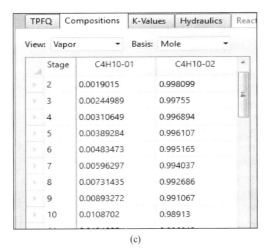

Figure 5.7 Viewing the temperature, pressure, flow rate and composition of the top reflux

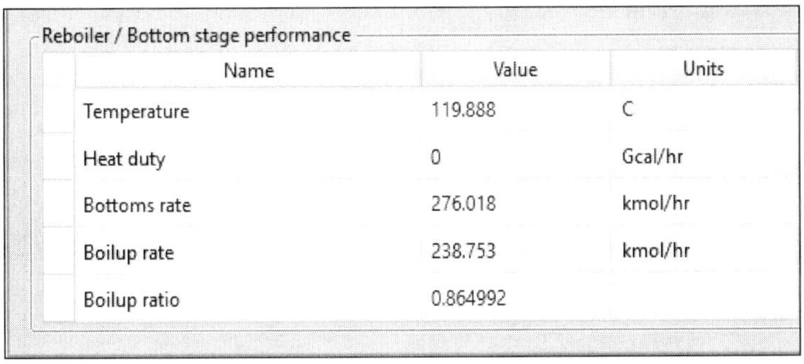

Figure 5.8 Viewing temperature, pressure, flow rate and composition of the return column stream

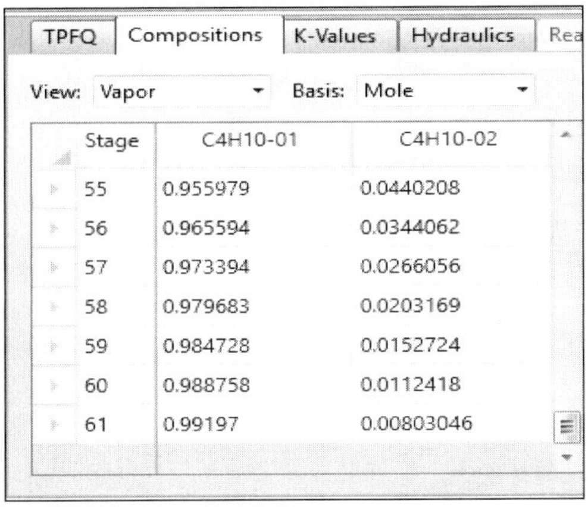

(3) Click **Next** to enter the **Streams| REB-OUT| Input| Mixed** page, input the pressure of REB-OUT as 0.11 MPa, gas phase fraction is 1, flow rate is 17696.6 kg/h, *n*-butanol mass fraction is 0.005, isobutanol mass fraction is 0.995.

(4) Click **Next** to enter the **Streams| REFLUX| Input| Mixed** page, input the REFLUX pressure as 0.6 MPa, gas phase fraction is 0, flow rate is 15935.7 kg/h, *n*-butanol mass fraction is 0.995, isobutanol mass fraction is 0.005.

(5) Click **Next** to enter the **Blocks | HPD | Specifications | Setup| Configuration** page and enter the heat pump distillation column parameters, as shown in Figure 5.9.

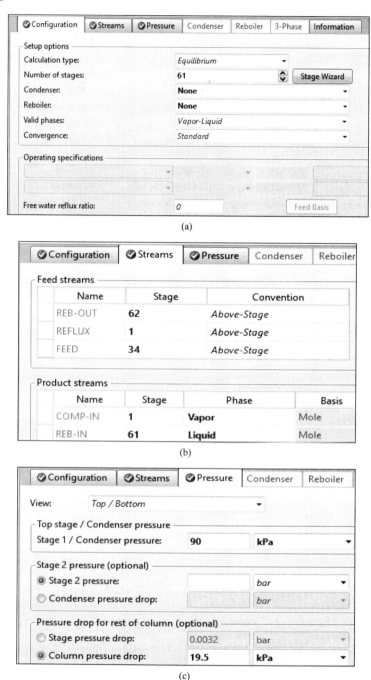

Figure 5.9 Inputting heat pump distillation column HPD parameters

(6) Add compressor, reboiler and splitter. Compressor COMP uses the **Pressure Changers |Comp| ICON2** icon in the module palette. Reboiler uses the **Exchangers| HeatX| GEN-HS** icon in the module palette. Splitter S1 uses **Mixers/Splitters|**

FSplit| TRIANGLE in the module palette, as shown in Figure 5.10.

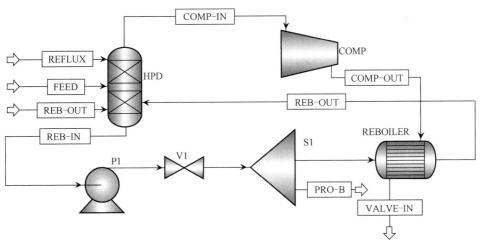

Figure 5.10　Adding compressor, reboiler and splitter

(7) Click **Next** to enter the **Blocks| COMP| Setup| Specifications** page and enter the COMP parameters of the compressor. The compressor type is ASMR multi-variable compression, the variable efficiency is 0.8, the mechanical efficiency is 0.95 and the compression ratio is 2.3, as shown in Figure 5.11.

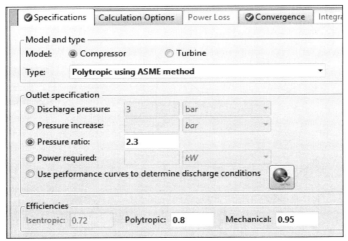

Figure 5.11　Inputting compressor COMP parameters

(8) Click **Next** to enter the **Blocks| REBOILER| Setup| Specifications** page, enter the cold stream outlet vapor fraction as 1, as shown in Figure 5.12.

(9) Add the auxiliary condenser and the splitter. The auxiliary condenser COOLER uses the **Exchangers| Heater | HEATER** icon in the module palette. The splitter S2 uses the **Mixers/Splitters| FSplit | TRIANGLE** icon in the module palette , as shown in Figure 5.13.

(10) Click **Next** to enter the **Blocks| COOLER| Input| Specifications** page and enter the auxiliary condenser COOLER parameters, as shown in Figure 5.14.

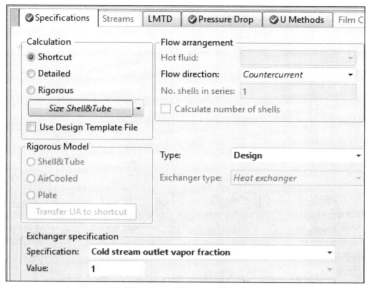

Figure 5.12 Inputting reboiler parameter

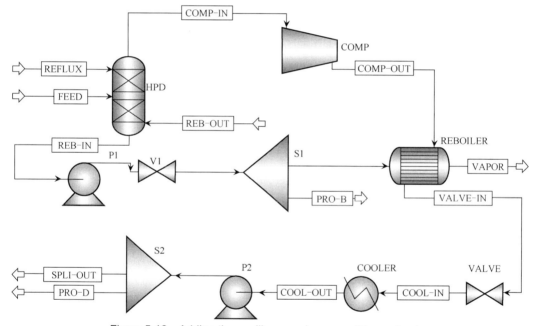

Figure 5.13 Adding the auxiliary condenser, splitter and valve

(11) Click **Next** to enter the **Blocks| SPLITTER| Input| Specifications** page and enter the SPLITTER parameter, as shown in Figure 5.15.

(12) Click **Next**, the Required Input Complete dialog box appears, click **OK**, run the simulation and the process converges.

Compare the input value of the logistics REFLUX with the calculated value of the logistics SPLI-OUT, calculate the value of the logistics SPLI-OUT as the input value of the logistics REFLUX, run the simulation until the two are close and merge the two streams.

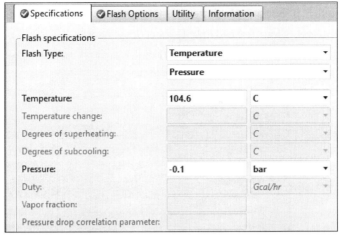

Figure 5.14　Inputting auxiliary condenser COOLER parameters

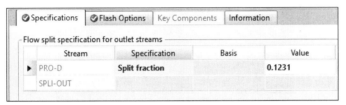

Figure 5.15　Inputting the SPLITTER parameter

Compare the input value of the logistics REB-OUT with the calculated value of the logistics VAPOR, the calculated value of the logistics VAPOR is used as the input value of the logistics REB-OUT and the simulation is running until the two are close to each other and the two streams are combined. Select logistics REB-OUT and logistics VAPOR, right click on a list, click **Join Streams**, merge two streams, as shown in Figure 5.16.

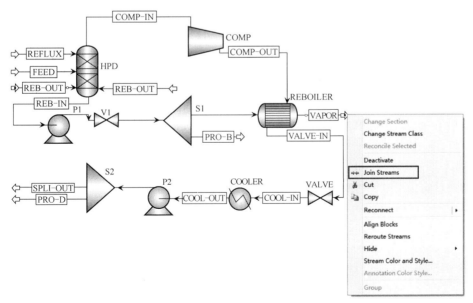

Figure 5.16　Consolidated logistics REB-OUT and logistics VAPOR

Go to the **Results Summary| Streams| Material** page and view the logistics results. The *n*-butanol mass fraction of the top stream is 0.998 and the isobutanol mass fraction of the bottom stream is 0.998, which meets the product purity requirements, as shown in Figure 5.17.

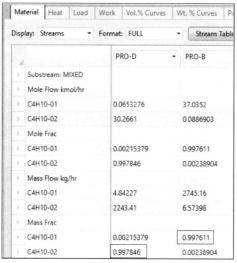

Figure 5.17　Viewing logistics results

(13) Go to the **Blocks| COMPR| Results| Summary** page, **Blocks| REBOILER| Results| Summary** page, **Blocks| COOLER| Results| Summary** page to view the compressor, auxiliary reboiler, and auxiliary condenser duty. The compressor consumes 257.961 kW, the auxiliary reboiler duty is 2842.68 kW, and the auxiliary condenser duty is -304.889 kW.

Exercises

1. The fresh feed is a liquid mixture containing 30% (mass fraction) isopropanol and 70% chlorobenzene at a mass flow rate of 5000 kg/h. The UNIQUAC property model was chosen as the thermodynamic model. The required products purity for the isopropanol and chlorobenzene is 99.8% (mass fraction) or above. Specific process operation parameters are shown in Figure 5.18. Conduct the simulation. (The case is from Liu. Heat pump distillation process and dynamics for close-boiling and special wide-boiling mixture [D]. Qingdao: Qingdao University of Science and Techenology, 2017.)

2. An existing propylene distillation column is used to separate propylene and propane, requiring the mole fraction of propylene at the top of the column not less than 99.6% and propane at the bottom of the column not less than 97.5%. Please use conventional distillation column and top steam direct compression heat pump distillation column to separate propylene and propane. The specific operating parameters are as follows.

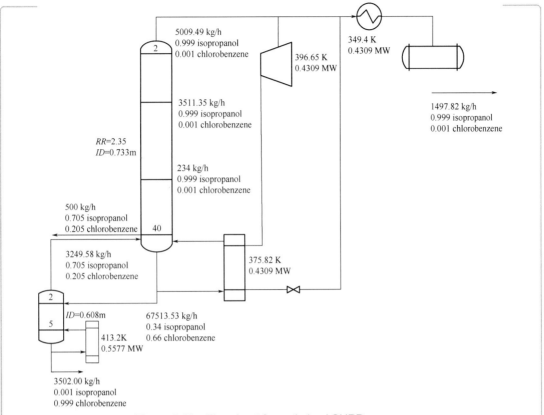

Figure 5.18 Flowsheet for optimized SHPD process

Bubble point feeding, pressure 2MPa. Feed flow rate is 100 kmol/h, in which propylene and propane mole fraction are 50%. The number of trays in conventional distillation column is 184, feed position is 140, reflux ratio is 14.975, product flow rate at bottom is 51.08 kmol/h, condenser pressure is 1.15 MPa, tray pressure drop is 0.0008 MPa, and physical property method is chosen as PENG-ROB.

References

[1] Gao X, Chen J, Ma Z, et al. Simulation and optimization of distillation processes for separating a close-boiling mixture of n-butanol and isobutanol[J]. Industrial & Engineering Chemistry Research, 2014, 53(37): 14440-14445.

[2] Fu C, King C, Chang Y, et al. Vapor-liquid and liquid-liquid phase equilibria of n-butanol-isobutanol-water system[J]. Chem Ind Eng, 1979, 5: 97-106.

[3] Liu X. Heat pump distillation process and dynamics for close-boiling and special wide-boiling mixture[D]. Qingdao: Qingdao University of Science and Techenology, 2017.

[4] Muhrer C A, Collura M A, Luyben W L. Control of vapor recompression distillation columns[J]. Industrial & Engineering Chemistry Research, 1990, 29(1):59-71.

[5] Karami G, Amidpour M, Sheibani B H, et al. Distillation column controllability analysis through heat pump integration[J]. Chemical Engineering and Processing: Process Intensification, 2015, 97: 23-37.

[6] Sun L. Chemical process simulation training - Aspen course [M]. Beijing: Chemical Industry Press, 2017.

Chapter 6

Energy-saving Side-stream Extractive Distillation Process

6.1 Introduction

The principle of extractive distillation technology is to add a third component (solvent) to the mixture to be separated to increase the relative volatility of the original components, so as to achieve effective separation of the mixture. The concept of extractive distillation was proposed by Dunn et al in 1945. Subsequently, the technology developed rapidly and was widely used in industry. Although extractive distillation technology is widely used in industry, the main obstacle is its high energy consumption. It is important to improve the energy efficiency of extractive distillation. Besides energy-saving technology, its dynamic control strategy is also an important factor to be considered in chemical production process.

Extractive distillation energy-saving technology includes heat-integration technology, heat pump technology and other technologies. It is found that these technologies can effectively reduce the annual total cost of the process. In 2016, Salvador et al proposed a new side-stream extractive distillation process. The results show that the side-stream extractive distillation process can reduce energy consumption more effectively than the thermally coupled extractive distillation process. Therefore, it is significant to apply the idea of side-flow connection to design new process and improve its dynamic control performance.

6.2 Steady-state Design of Side-stream Extractive Distillation

In this work, the steady-state and dynamics of this process are simulated using Aspen Plus and Aspen Dynamics software. The steady-state design of the side-stream extractive distillation for separating methanol and acetone refers to the work by Salvador's paper. The flowsheet is shown in Figure 6.1. The feed flow rate is set to 540 kmol/h with the composition of 50% methanol and 50% acetone (mole fraction). The purity of product is 99.5% (mole fraction) for methanol and acetone.

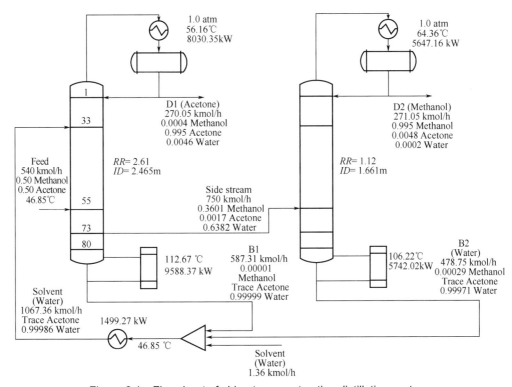

Figure 6.1 Flowsheet of side-stream extractive distillation system

6.3 Dynamic Control of Side-stream Extractive Distillation

Accurate control structures for the side-stream distillation process are necessary. In this work, we explore the detailed control strategies of this energy saving system for separating methanol and acetone using water as solvent. The ±10% and ±20% feed flow rate and composition disturbances are introduced to test the stability and robustness of the control strategies.

Before exporting from Aspen Plus software to the Aspen dynamic mode, the main parameter should be set as follow. The tray pressure drop of the two processes is set to 0.0068 atm. When the vessel is half full, there is a 5 min liquid holdup. It is very necessary that valid phases of all valves are set to Liquid-Only. Figure 6.2 shows temperature and temperature slope profiles of side-stream extractive distillation.

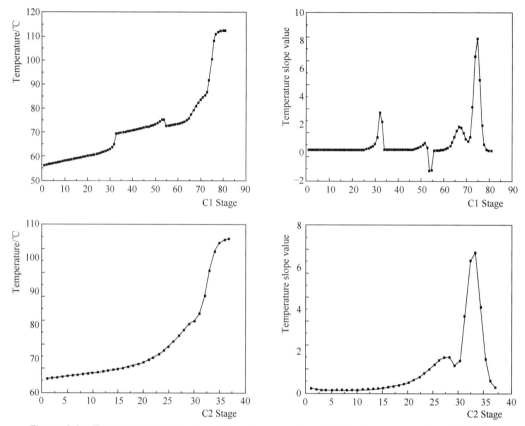

Figure 6.2　Temperature and temperature slope profiles of side-stream extractive distillation

As shown in Figure 6.3, the original control structure of the side-stream extractive distillation process is proposed, based on the conventional control structure of the extractive distillation system for separating binary mixture of acetone and methanol. The detailed control structures and the related settings are listed below:

(1) The fresh feed flow is controlled by throughput valve (reverse acting).

(2) The distillate flow rates are manipulated to control the reflux tank levels in two columns (direct acting).

(3) The makeup water flow rate is manipulated to hold the sump level of the extractive distillation column C1 (reverse acting).

(4) The bottom rate of solvent recovery distillation column is manipulated to hold the sump level of column C2 (direct acting).

(5) The operating pressures in two columns are controlled by manipulating the condenser duties of two columns (reverse acting).

(6) The cooler heat duty is manipulated to control the temperature of recycling solvent (reverse acting).

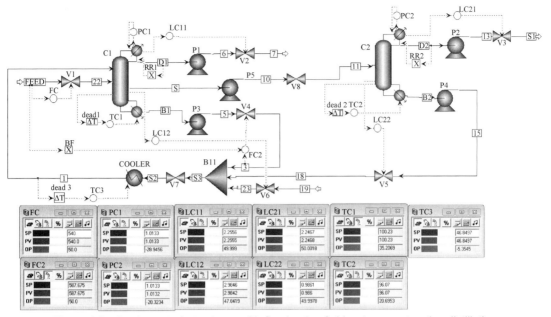

Figure 6.3 Basic control structures with fixed ratio of side-stream extractive distillation

(7) The solvent flow rate at the bottom of extractive distillation column C1 is directly proportion to the total feed flow rate.

(8) The temperature in stage 75 of C1 column and the temperature in stage 33 of C2 column are controlled by operating the corresponding reboiler duty input (reverse acting).

As for basic side-stream extractive distillation controller tuning parameters, level control is set with large integral time (9999 min) and $K_c=2$ to implement proporation-only control. For pressure controllers in both columns, the tuning parameters are set at $K_c = 20$ and $\tau_I = 12$ min. The total feed flow controller settings are $K_c = 0.5$ and $\tau_I = 0.3$ min. Three deadtime elements (1 min) are inserted into three corresponding temperature control loops. Then relay-feedback tests are run, and the Tyreus–Luyben tuning rule is used to determine tuning parameters of three temperature controllers.

The dynamic performances of basic control structures are tested when the flow rate disturbances and feed composition disturbances are introduced. As shown in Figure 6.4, the flow rate disturbances and the feed composition disturbances are introduced at 0.5 h, and the process is paused at 20 h. It is noticed that product purity of two columns is not able to return to the desired values at the new steady-state when $\pm 10\%$ feed disturbances are introduced. Through the analysis of basic control in dynamic process, acetone was delivered into the C2 column by side-stream. It is impossible for methanol to be purified in C2 column. Therefore, some improved control schemes should be further explored to achieve efficient control.

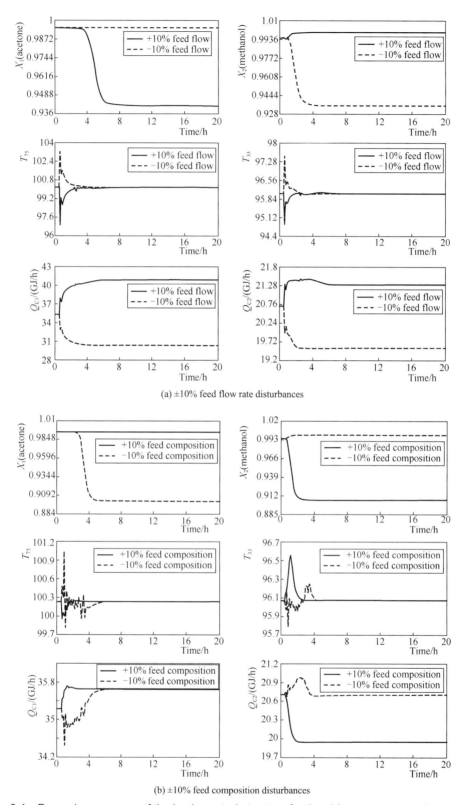

Figure 6.4 Dynamic responses of the basic control structure for the side-stream extractive distillation

6.3.1 Control Structure with Side-stream Composition/Temperature Cascade Connection

On the basis of basic control structure, a cascade side-stream composition/temperature control structure was introduced. The control structure is shown in Figure 6.5. The acetone purity in the side-stream is detected (3 min deadtime) and controlled by manipulating the set point of the temperature controller which is on cascade. This control structure is explored to control the acetone purity in the side-stream. ±10% disturbances are introduced and the control performances are shown in Figure 6.6. As can be seen, this structure can be kept constant after 4 h and the acetone purity is well held under the ±10% feed disturbances. However, the methanol purity can reach 99.41% instead of 99.5% (mole fraction) after introducing ±10% feed disturbances. This control structure can be selected when the purity requirement of methanol is not high.

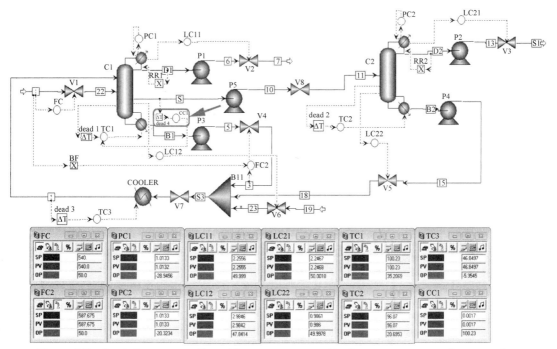

Figure 6.5 Control structure with a side-stream composition/temperature cascade connection

6.3.2 Control Structure with S/F and Composition Controller Connection

To achieve better dynamic controllability, finding the appropriate controller to manipulate side-stream is the key factor. Professor William L Luyben reviewed and proposed the control structure of side-stream extractive distillation with side-stream flow rate/feed flow rate ratio(S/F) and composition controller connection. S/F control structure and composition controller connection is added to control the flow rate of the side-stream based on the basic control structure.

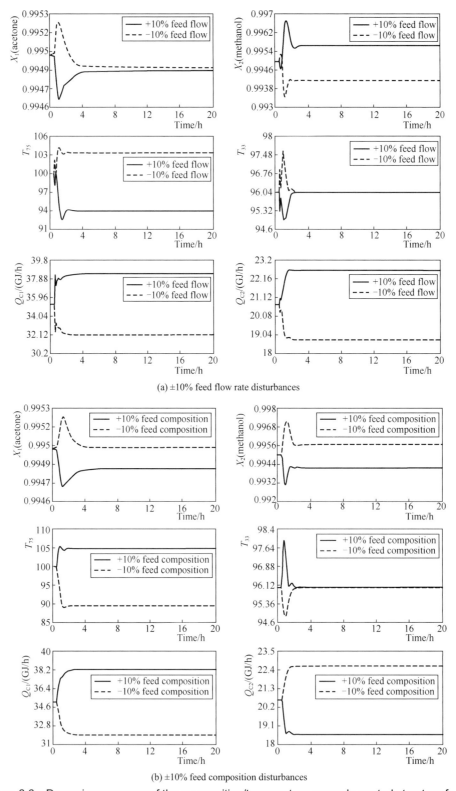

Figure 6.6 Dynamic responses of the composition/temperature cascade control structure for the side-stream extractive distillation

The methanol mole fraction signal is fed as the process variable into a composition controller whose output signal changed the S/F ratio. The control structure is shown in Figure 6.7. To test the control performance, ±10% feed disturbances are introduced. The dynamic performance results are shown in Figure 6.8 and reveal that the control structure can effectively control both processes disturbances. Meanwhile, the control structure can be kept constant after 8 h under the ±10% flow rate disturbances, and the system takes about 16 h to return to the desired value under the ±10% feed composition disturbances. The ±20% feed flow rate and feed composition disturbances are introduced and the performance results are shown in Figure 6.9. The results indicate that the system cannot reach a new steady-state within 20 h after ±20% feed composition disturbances were introduced.

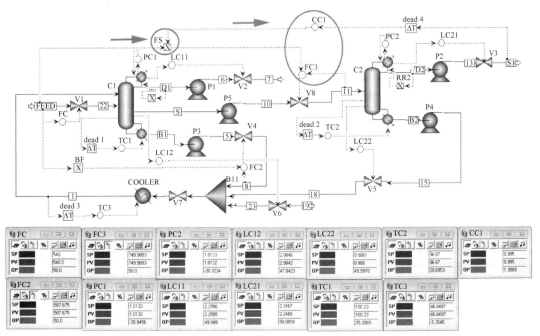

Figure 6.7　Control structure with side-stream flow rate/feed flow rate ratio(S/F) and composition controller connection

6.3.3　Improved Dynamic Control Structure

Through the investigation of the above three structures, the results indicate that a suitable side-stream flow rate control structure is crucial to obtain effective control of side-stream extractive distillation. A new controller structure is proposed to control side-stream flow rate and composition in this process. The difference between the basic and improved side-stream extractive distillation control structure is that the product purity of solvent recovery column C2 is controlled by manipulating side-stream flow rate (direct acting). The improved control structure is shown in Figure 6.10.

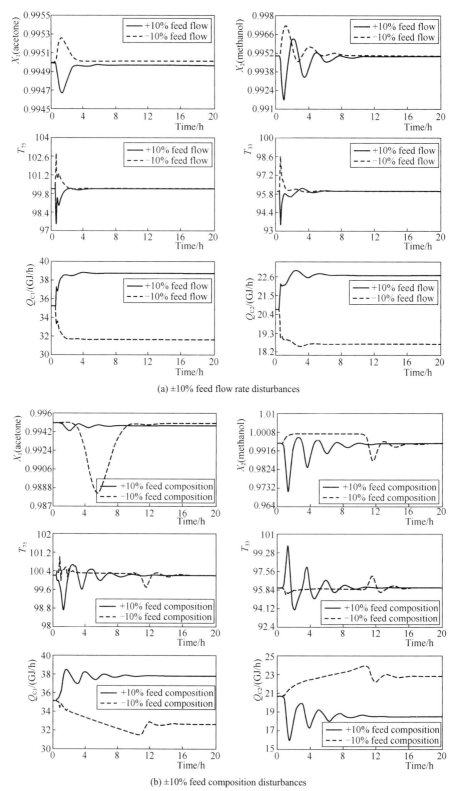

(a) ±10% feed flow rate disturbances

(b) ±10% feed composition disturbances

Figure 6.8　Dynamic responses of control structure with S/F and composition cascade connection

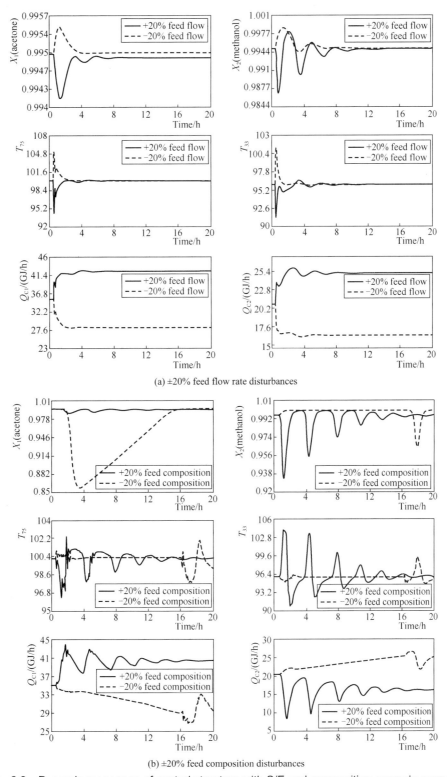

(a) ±20% feed flow rate disturbances

(b) ±20% feed composition disturbances

Figure 6.9 Dynamic responses of control structure with S/F and composition cascade connection

For controller tuning parameters, the setting is the same as that of basic dynamic process. Figure 6.11 shows the dynamic performances of the improved control strategies under ±10% flow rate disturbances and ±10% feed composition disturbances. It is noticed that product purity can come very close to the initial values at the new steady-state about 8 h when ±10% flow rate disturbances were introduced. This structure can also keep constant and reach the desired values after 12 h for ±10% feed composition disturbances. Therefore, the whole control structure can achieve good controllability for side-stream extractive distillation.

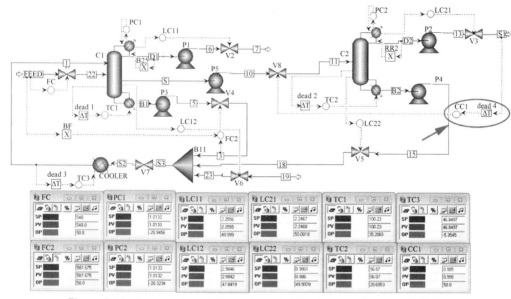

Figure 6.10 Improved control structure of side-stream extractive distillation

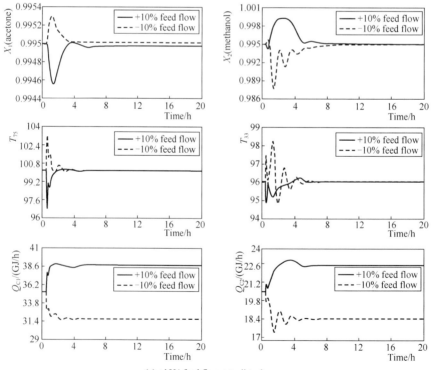

(a) ±10% feed flow rate disturbances

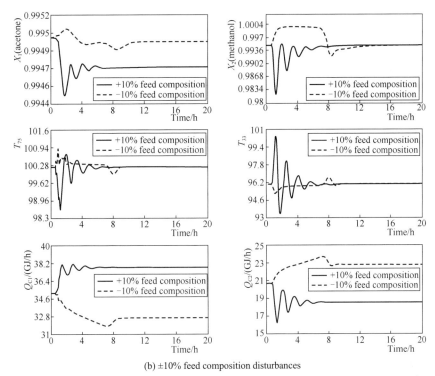

(b) ±10% feed composition disturbances

Figure 6.11 Dynamic responses of the improved control structure for the side-stream extractive distillation

Exercises

1. The side-stream extractive distillation process for separation of water and ethanol system was designed. Detailed information about the steady-state process is shown in Figure 6.12. Complete the whole process.

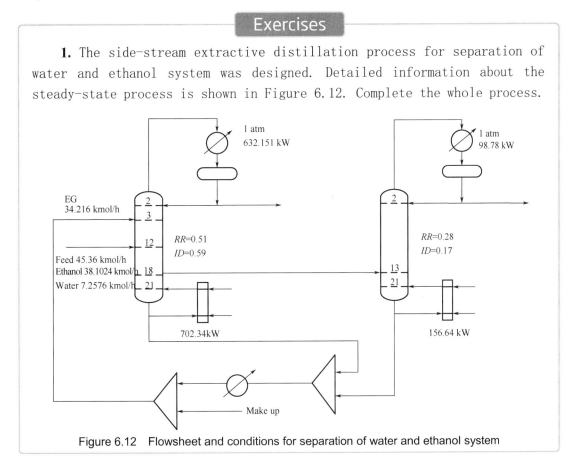

Figure 6.12 Flowsheet and conditions for separation of water and ethanol system

2. The side-stream extractive distillation process for separation of acetone and methanol system was designed. Detailed information about the steady-state process is shown in Figure 6.13. Complete the whole process.

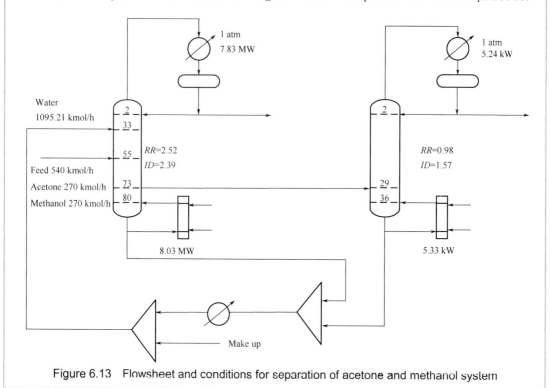

Figure 6.13 Flowsheet and conditions for separation of acetone and methanol system

References

[1] Tututi-Avila S, Medina-Herrera N, Hahn J, et al. Design of an energy-efficient side-stream extractive distillation system [J]. Computers & Chemical Engineering, 2017, 102: 17-25.

[2] Ma K, Yu M, Dai Y, et al. Control of an energy-saving side-stream extractive distillation process with different disturbance conditions [J]. Separation and Purification Technology, 2019, 210: 195-208.

[3] Luyben W L. Control comparison of conventional and thermally coupled ternary extractive distillation processes [J]. Chemical Engineering Research and Design, 2016, 106: 253-262.

[4] Ma Y, Cui P, Wang Y, et al. A review of extractive distillation from an azeotropic phenomenon for dynamic control [J]. Chinese Journal of Chemical Engineering, 2018.

Chapter 7

Pressure-swing Distillation for Minimum-boiling Azeotropes

7.1 Introduction

Separation of azeotrope is a common problem in chemical and pharmaceutical industry. If the azeotrope system is sensitive to pressure, the azeotrope can be separated by changing the pressure, and the pressure-swing distillation has obvious economic advantages compared with other distillation processes. Pressure-swing distillation can be used to separate the maximum-boiling azeotrope, minimum-boiling azeotrope and the azeotrope with close boiling point according to the characteristics of azeotrope.

In this chapter, an effective control structure is designed for the separation of n-heptane and isobutanol by pressure-swing distillation. Based on this control structure, the process control structure with full heat-integration is further explored.

7.2 Converting from Steady-state to Dynamic Simulation

Figure 7.1 shows the process for separating n-heptane and isobutanol by pressure-swing distillation without heat-integration. Figure 7.2 shows the process for separating n-heptane and isobutanol by pressure-swing distillation with full heat-integration. This section will develop the control structures based on these processes.

变压精馏的动态控制

Before the steady-state simulation file is imported into the dynamic simulation file, it is necessary to set the equipment size of the distillation column, the pressure drop of the distillation column, the pressure difference of the inlet and

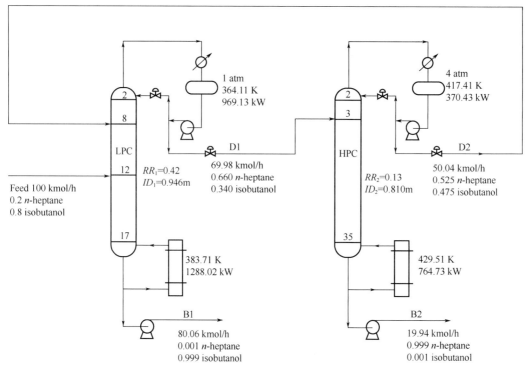

Figure 7.1 Flowsheet of the pressure-swing distillation without heat-integration

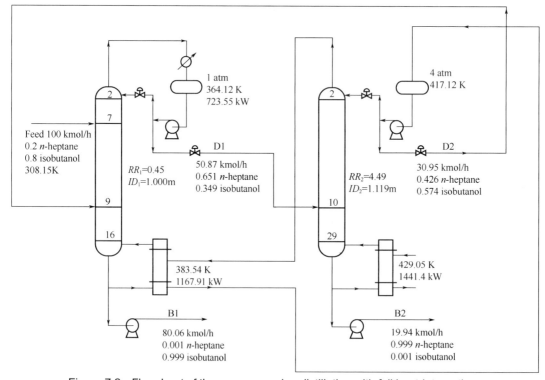

Figure 7.2 Flowsheet of the pressure-swing distillation with full heat-integration

outlet of the pump and the pressure drop of the valve, etc. It should be noted that the sizes of the reflux drum and kettle in the distillation column should be determined as follows. It takes 5 minutes for the equipment to reach its half liquid capacity, and the ratio of height to diameter is 2. Table 7.1 shows the parameters of equipments in the nonheat-integration and full heat-integration processes.

Table 7.1 Parameters of equipments in the nonheat-integration and full heat-integration processes

Parameters	Nonheat-integration		Full heat-integration	
	LPC	HPC	LPC	HPC
Flow in drump / (m³/s)	0.003882	0.002292	0.002870	0.006562
Volume of drump / m³	2.33	1.38	1.72	3.94
Diameter of drump / m	1.14	0.96	1.03	1.36
Length of drump / m	2.28	1.92	2.06	2.72
Flow in sump / (m³/s)	0.005546	0.006149	0.005241	0.010674
Volume of sump / m³	3.33	3.69	3.14	6.40
Diameter of sump / m	1.28	1.33	1.26	1.60
Length of sump / m	2.56	2.66	2.52	3.20

The choice of temperature-sensitive stage has an important influence on the process control structure. The temperature detector detects the temperature change of the sensitive stage and transmits the signal to the temperature controller, which then transmits the signal to the actuator to change the operation variables to realize the process control.

The temperature and its slope profiles of the two columns in the nonheat-integration and full heat-integration processes are shown in Figure 7.3 and Figure

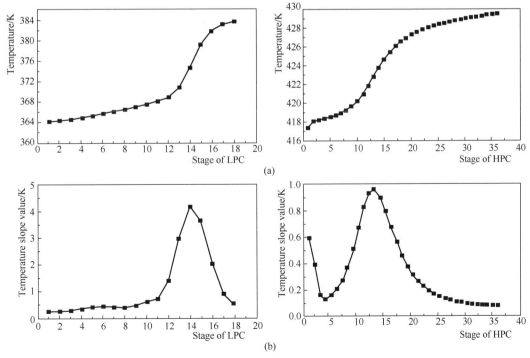

Figure 7.3 Temperature and its slope profiles of the LPC and HPC in the nonheat-integration process

7.4, respectively. It is found that the slope profile of each column has one peak and the corresponding stages (i.e., stage 14 and 13 in the nonheat-integration process and stage 13 and 12 in the full heat-integration process) are selected as control stages according to the slope criterion.

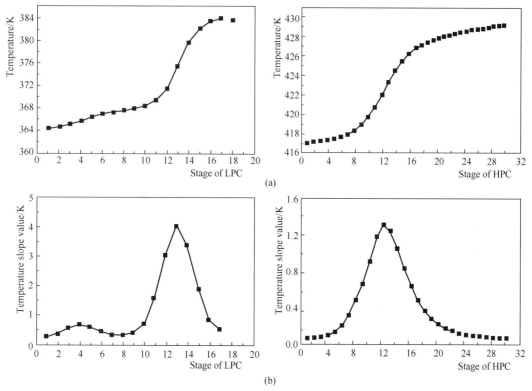

Figure 7.4   Temperature and its slope profiles of the LPC and HPC in the full heat-integration process

7.3   Control Structures of the Process without Heat-integration

7.3.1   Basic Control Structure

As shown in Figure 7.5, after importing the steady-state simulation into the dynamic simulation, open the dynamic simulation file.

The very first thing to do with any newly imported file is to make an "initialization" run to make sure that everything is running. At the very top of the screen there is a little window that says Dynamic. Clicking the arrow to the right opens the dropdown menu as shown in Figure 7.5a. Select **Initialization** and click the **Run** button, which is just to the upper right. If everything is set up correctly, the window shown in Figure 7.5b opens. The next thing to do is to make sure the integrator is working correctly. This is done by changing **Initialization mode** to **Dynamic** and clicking the **Run** button again. As shown in

Figure 7.5a, the block at the bottom of the screen is green. If something goes wrong, this block will turn red and you will not be able to run the simulation. The run is stopped by clicking the **Pause** button.

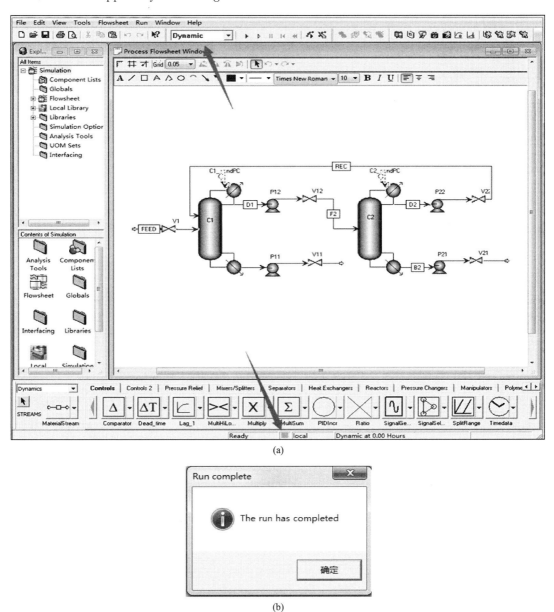

Figure 7.5 (a) Selecting the initialization; (b) Initialization run successfully

In the initial flowsheet, the pressure controller is the default controller. In the basic control structure, the liquid level controller, flow controller and temperature controller are installed. First, the liquid level controller is added, click **Simulation | Libraries | Dynamics | Control Models** and select **PIDIncr**, or select it from the model library below, as shown in Figure 7.6. Drag it onto the pump P11, P12, P21, P22, and name LC11, LC12, LC21, LC22.

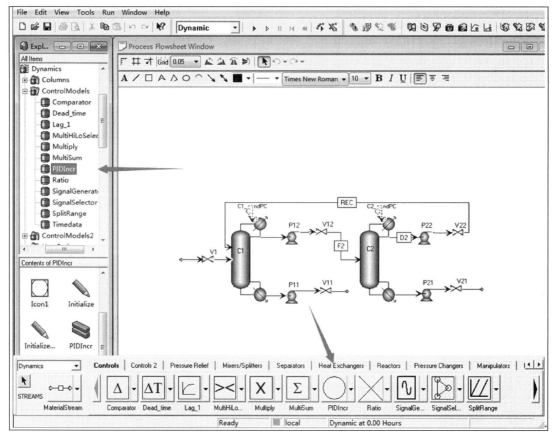

Figure 7.6 Selecting the PIDIncr

Next, connect the control signal. First select control signal, draw the arrow from the condenser, select the liquid level of the first stage, click **OK** to connect the control signal line to the condenser, as shown in Figure 7.7a and connect the other end to the controller. Clicking on the blue arrow on the left side of the controller opens the window, as shown in Figure 7.7b. Clicking **OK** completes the control signal connection between the condenser and the level controller. Then connect the signal to the valve, as shown in Figure 7.7c.

Finally, set the controller parameters, double-click the controller LC12 to open the control panel, and then click **Configure** to enter the tuning interface, as shown in Figure 7.8. The action of the controller should be direct because if the level increases, the signal to the valve should increase. The controller gain is set to 2 and the integral time is set at a very large number (9999 min). Click **Initialize Values** to initialize the data.

The process of installing a level controller on the base of the column is similar to the process mentioned above, but it should be noted that the liquid level of the last stage is selected in the interface of connecting control signal, as shown in Figure 7.9.

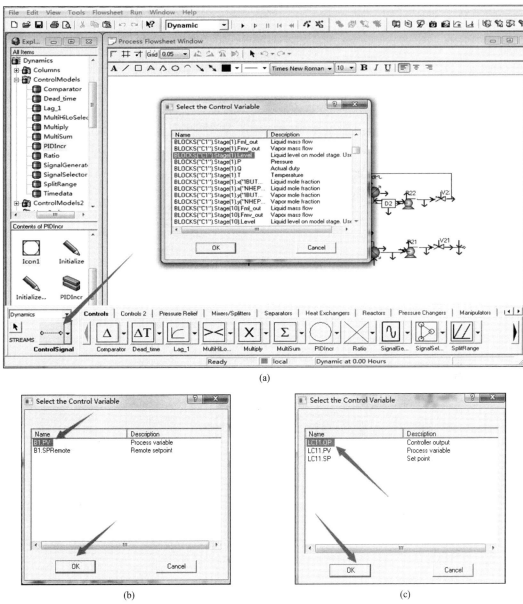

Figure 7.7 (a) Connecting control signal; (b) Connecting the signal to the controller;
(c) Connecting the signal to the valve

The process of installing a flow controller is similar to the process of installing a level controller, but the option selected in the interface of connecting control signal and the data in tuning interface are different, as shown in Figure 7.10.

Installing temperature controller is somewhat more involved than installing level and flow controller. The process settings are shown in Figure 7.11. The temperature of the 14th stage is chosen because the 14th stage is the temperature-sensitive stage. OP is selected as the reboiler heat input **QRebR**, the controller action is **Reverse**.

Figure 7.8 (a) The control panel; (b) The tuning interface

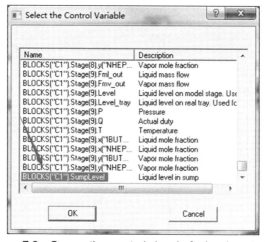

Figure 7.9 Connecting control signal of a level controller

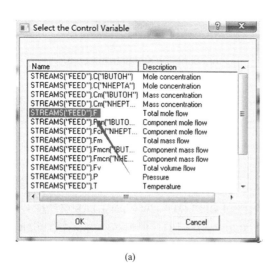
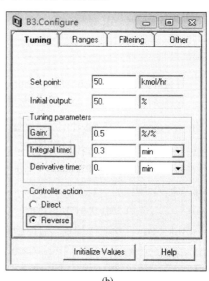

(a) (b)

Figure 7.10 (a) Connecting control signal; (b) The tuning interface

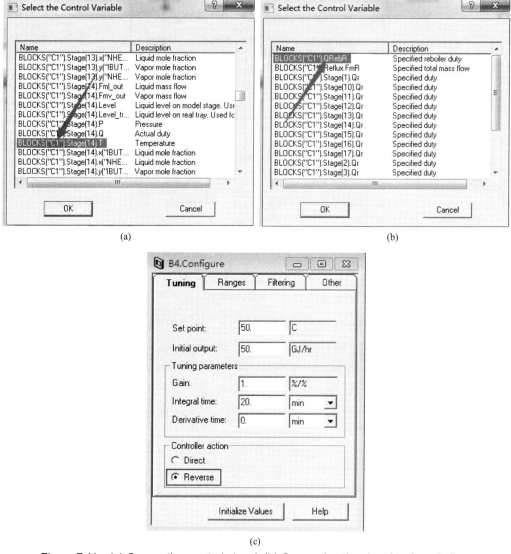

Figure 7.11 (a) Connecting control signal; (b) Connecting the signal to the reboiler; (c) The tuning interface

Next, insert a deadtime element on the flowsheet between the column and the TC1 temperature controller. The reason for installing the controller initially without the deadtime element is to avoid initialization problems. The line from the temperature of the 14th stage is selected. Right-clicking, selecting **Reconnect Destination,** and placing the icon on the arrow pointing to the deadtime icon, the input for the deadtime is connected. A new control signal is inserted between the deadtime and the controller. The deadtime icon is then selected. Right-clicking, selecting **Forms** from the dropdown list, and selecting **All Variables,** open the window, as shown in Figure 7.12a. The DeadTime is 0 min initially. A deadtime of 1 min is entered, and performing an initialization run fills in the correct values, as shown in Figure 7.12b.

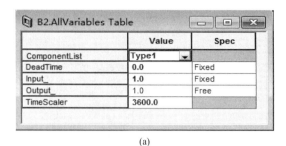

(a)

(b)

Figure 7.12　AllVariables table (a) before initialization running; (b) after initialization running

Everything is ready for the relay-feedback test. Clicking the **Tune** button, as shown in Figure 7.13a. Specify a **Closed loop ATV** as the Test method. To start the test, click the **Run** button at the top of the screen and click the **Start test** button on the Tune window. To be able to see the dynamic responses, click the **Plot** button at the top of the controller faceplate. After several (4~6) cycles have occurred, click the **Finish test** button. Figure 7.13b gives the results. Finally, the **Tuning parameters** page tab is clicked, the **Calculate** button is pushed. The resulting controller settings are gain=1.47 and integral time=11.88 min. These are loaded into the controller by clicking the **Update controller** button, as shown in Figure 7.13c.

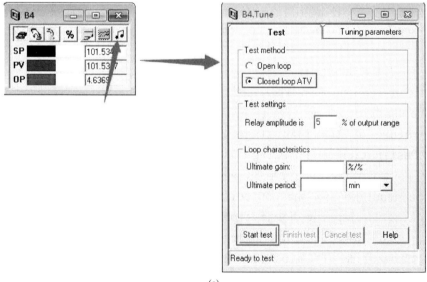

(a)

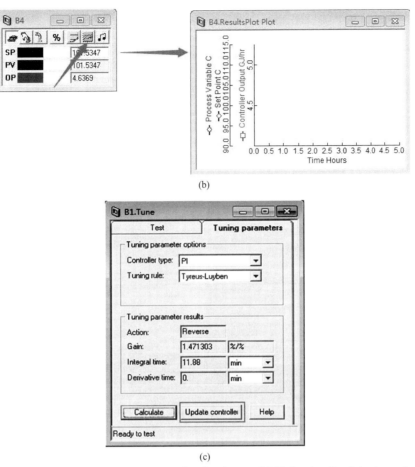

Figure 7.13 (a) Setting up the relay-feedback test; (b) Relay-feedback test results;
(c) Calculated controller settings

The R/F structure is implemented by using a multiplier block. The input of this block is the mass flow rate of the feed, which is 7933.9 kg/h. Figure 7.14 gives the stream information for the feed F1. The output of the multiplier block is the mass flow rate of the reflux. To determine the design value of this variable, click the **column** icon, right-click, select **Forms** and **Results**. This opens the table shown in Figure 7.15a. The reflux mass flow rate is 3332.23 kg/h. So the multiplier block should multiply the feed mass flow rate by the number (3332.23/7933.9) = 0.4199.

A multiplier control model is placed on the flowsheet. The control signal is connected from the feed to the multiplier R/F. Another control signal is connected from the multiplier output to the column (the blue arrow pointing to the line below the condenser). A list of alternatives opens, and the top one, **Reflux.FmR,** is selected, as shown in Figure 7.15b. To set the constant in the multiplier, the icon is clicked and right-clicked. **Forms** and **All Variables** are selected, and the window shown in Figure 7.15c opens, on which the number 0.4199 is entered for Input2.

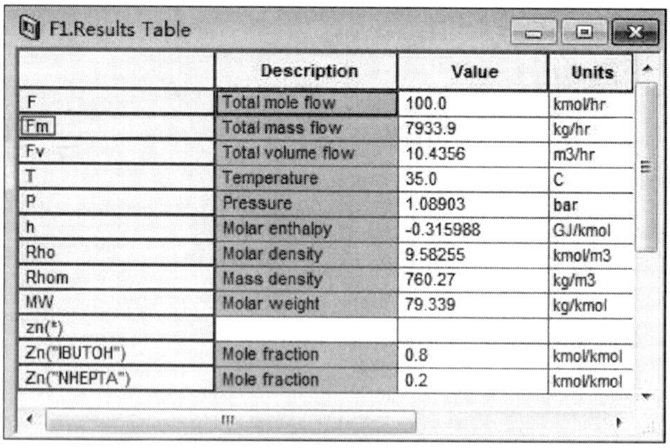

F1.Results Table

	Description	Value	Units
F	Total mole flow	100.0	kmol/hr
Fm	Total mass flow	7933.9	kg/hr
Fv	Total volume flow	10.4356	m3/hr
T	Temperature	35.0	C
P	Pressure	1.08903	bar
h	Molar enthalpy	-0.315988	GJ/kmol
Rho	Molar density	9.58255	kmol/m3
Rhom	Mass density	760.27	kg/m3
MW	Molar weight	79.339	kg/kmol
zn(*)			
Zn("IBUTOH")	Mole fraction	0.8	kmol/kmol
Zn("NHEPTA")	Mole fraction	0.2	kmol/kmol

Figure 7.14　Results table for feed F1

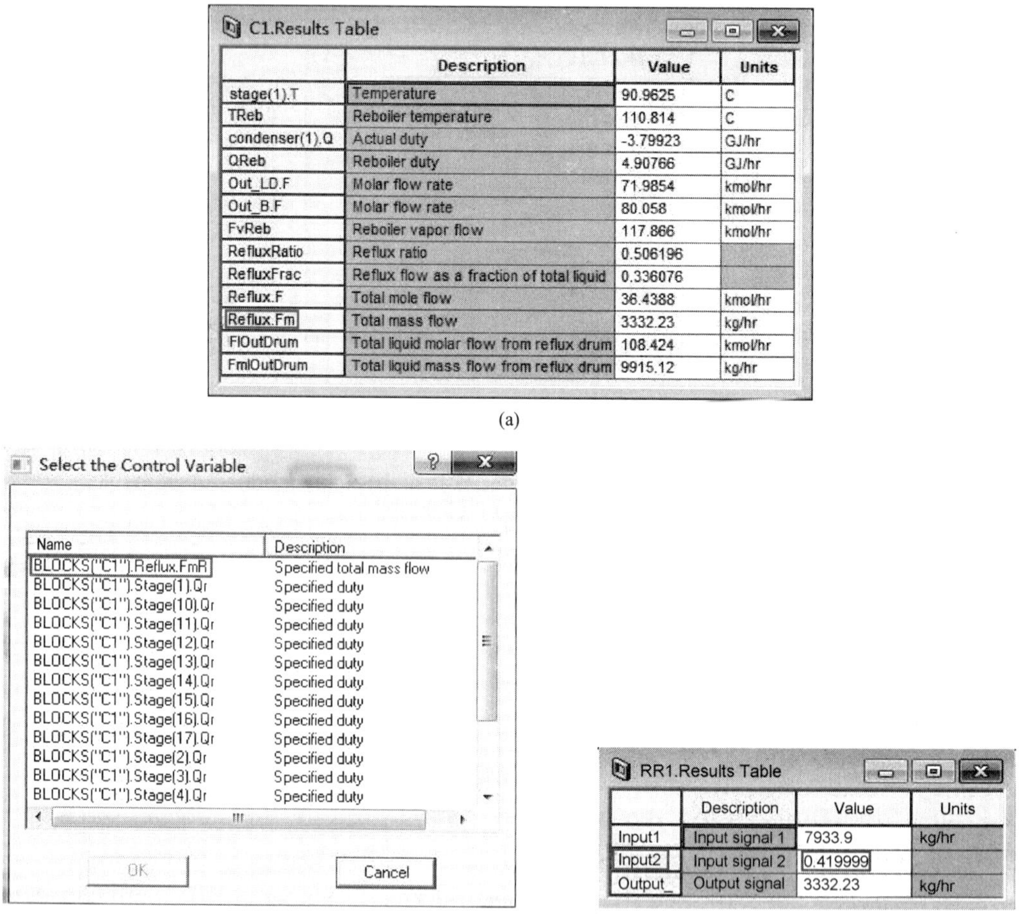

C1.Results Table

	Description	Value	Units
stage(1).T	Temperature	90.9625	C
TReb	Reboiler temperature	110.814	C
condenser(1).Q	Actual duty	-3.79923	GJ/hr
QReb	Reboiler duty	4.90766	GJ/hr
Out_LD.F	Molar flow rate	71.9854	kmol/hr
Out_B.F	Molar flow rate	80.058	kmol/hr
FvReb	Reboiler vapor flow	117.866	kmol/hr
RefluxRatio	Reflux ratio	0.506196	
RefluxFrac	Reflux flow as a fraction of total liquid	0.336076	
Reflux.F	Total mole flow	36.4388	kmol/hr
Reflux.Fm	Total mass flow	3332.23	kg/hr
FlOutDrum	Total liquid molar flow from reflux drum	108.424	kmol/hr
FmlOutDrum	Total liquid mass flow from reflux drum	9915.12	kg/hr

(a)

Select the Control Variable

Name	Description
BLOCKS("C1").Reflux.FmR	Specified total mass flow
BLOCKS("C1").Stage(1).Qr	Specified duty
BLOCKS("C1").Stage(10).Qr	Specified duty
BLOCKS("C1").Stage(11).Qr	Specified duty
BLOCKS("C1").Stage(12).Qr	Specified duty
BLOCKS("C1").Stage(13).Qr	Specified duty
BLOCKS("C1").Stage(14).Qr	Specified duty
BLOCKS("C1").Stage(15).Qr	Specified duty
BLOCKS("C1").Stage(16).Qr	Specified duty
BLOCKS("C1").Stage(17).Qr	Specified duty
BLOCKS("C1").Stage(2).Qr	Specified duty
BLOCKS("C1").Stage(3).Qr	Specified duty
BLOCKS("C1").Stage(4).Qr	Specified duty

OK　Cancel

(b)

RR1.Results Table

	Description	Value	Units
Input1	Input signal 1	7933.9	kg/hr
Input2	Input signal 2	0.419999	
Output_	Output signal	3332.23	kg/hr

(c)

Figure 7.15　(a) Results table for column; (b) Connecting the signal to the condenser; (c) Entering ratio constant

The final flowsheet and controller faceplates are shown in Figure 7.16.

In order to verify the performance of this control structure, a disturbance will be applied to it.

The first thing is to set up a plot, it will show how the variables of interest change dynamically with time. To open a plot, go to Tools on the top toolbar and

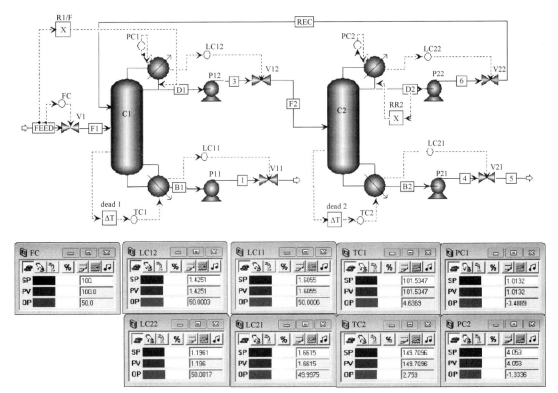

Figure 7.16 The final flowsheet and controller faceplates

click **New Plot**. As shown in Figure 7.17, a number of variables will be plotted: flow rate of the feed, the temperature of stage 14 and 13, reboiler heat input and bottoms impurity.

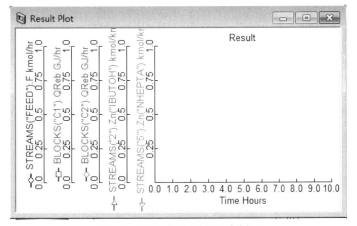

Figure 7.17 Plot with variables

Two disturbances will be applied to the control structure. Feed flow rate is changed by changing the feed flow controller setpoint. Feed composition is changed by clicking on the **Feed stream** icon, right-clicking, and selecting **Forms** and then **Manipulate**. Next, import the dynamic results into MATLAB to draw the disturbance result graph, as shown in Figure 7.18.

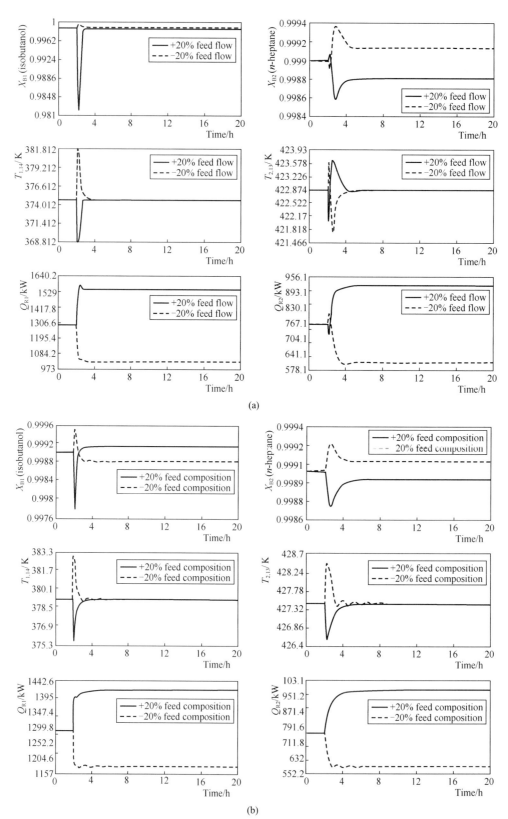

Figure 7.18 (a) Dynamic performance of the basic control structure under the feed flow rate disturbances; (b) Dynamic performance of the basic control structure under the feed composition disturbances (mole fraction)

7.3.2 Q_R/F_1 Control Structure

Aiming at the large instantaneous deviation of product purity of the two columns after encountering feed disturbance in the basic control structure, a proportional control module of heat load proportional to feed amount is added to the low pressure column in the improved control structure.

Select the Multiply module and name it Q_R/F_1 to connect the control signal, at this point, it is important to disconnect the output signal of the temperature controller, then connect the output signal of the temperature controller to Q_R/F_1 module, and connect the control signal to the reboiler. Before setting the parameters for Q_R/F_1 module, the unit of measure in Q_R/F_1 module must be set as GJ/kmol. The parameters are set by clicking on the Q_R/F_1 module, right-clicking, and selecting **Forms** and then **AllVariables**. The first input is the molar flow rate of the feed stream (100 kmol/h). The second input is the output signal of the temperature controller, and the output signal is the heat load of the reboiler (4.63687 GJ/hr). The second input should be (4.63687/100)=0.0463687, as shown in Figure 7.19. Then double-click the temperature controller, click **Configure**, and click **Initialize Values** to initialize and load the value into the temperature controller. The final flowsheet and controller faceplates is shown in Figure 7.20. Of course, the temperature controller must be retuned.

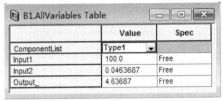

Figure 7.19　Entering the value of Q_R/F_1

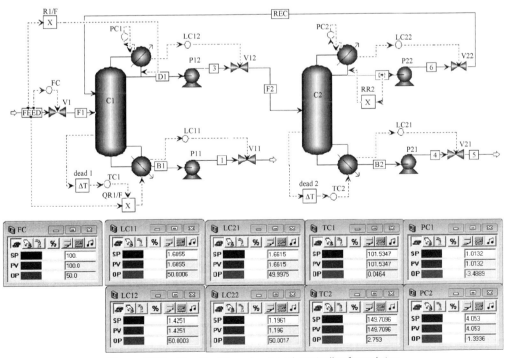

Figure 7.20　The final flowsheet and controller faceplates

Chapter 7　Pressure-swing Distillation for Minimum-boiling Azeotropes

7.3.3 Control Structures of the Process with FullHeat-integration

Through the "flowsheet" functions in "Constraints", the heat-integration of the pressure-swing distillation is realized in the dynamic simulation. Firstly, the reboiler area of the low pressure column is determined according to the heat load of the reboiler, the heat transfer temperature difference of the reboiler and the heat transfer coefficient of the reboiler. The heat load of the reboiler in the low pressure column is 4.20 GJ/h, and the heat transfer coefficient of the reboiler is 0.0020448 GJ/(h • m^2 • K). The heat transfer area of the reboiler in the low pressure column is 61.24 m^2. Figure 7.21 shows the complete heat-integration process formula written in the simulation. The first formula is to calculate the heat load of the low pressure column reboiler, and the second formula stipulates that the heat load of the condenser in the high pressure column and the heat load of reboiler in the low pressure column are equal in value.

```
Constraints - Flowsheet*
CONSTRAINTS
  // Flowsheet variables and equations...
   Blocks ("C1").QReb=61.24*0.0020448*(Blocks("C2").Stage(1).T-Blocks("C1").TReb);
   Blocks ("C2").condenser(1).Q=-Blocks("C1").QReb;
END
```

Figure 7.21 Flowsheet equations for the control stucture with fullheat-integration

Because of the fullheat-integration process, the heat of the low pressure column is provided by the high pressure column headgas phase, so the temperature of the sensitive stage of the high pressure column is maintained by controlling the heat load of the reboiler, and the temperature of the sensitive stage of the low pressure column is maintained by controlling the reflux ratio. The final flowsheet and controller faceplates is shown in Figure 7.22.

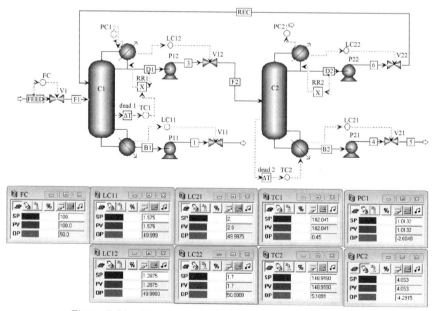

Figure 7.22 The final flowsheet and controller faceplates

Exercises

1. The process of separating ethanol-toluene azeotropic system via pressure-swing distillation with fullheat-integration was designed. Detailed information about the steady-state process is shown in Figure 7.23. Complete the dynamic control of the process. (The case is from Computers & Chemical Engineering, 2015, 76: 137-149.)

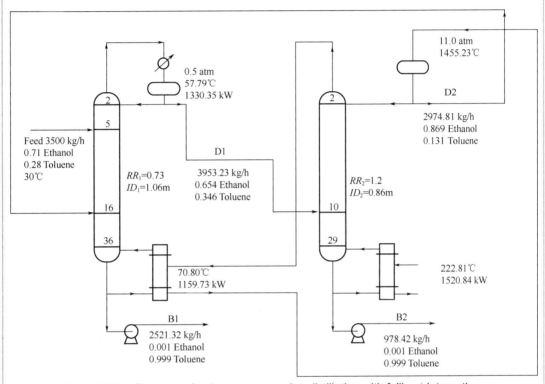

Figure 7.23　Flowsheet for the pressure-swing distillation with fullheat-integration

2. The process of the separation of acetonitrile/methanol/benzene ternary azeotrope via triple column pressure-swing distillation was designed. Detailed information about the steady-state process is shown in Figure 7.24. Complete the dynamic control of the process.

3. The process of the separation of the methanol-chloroform mixture process via pressure-swing distillation was designed. Detailed information about the steady-state process is shown in Figure 7.25. Complete the dynamic control of the process. (The case is from Computers & Chemical Engineering, 2014, 67: 166-177.)

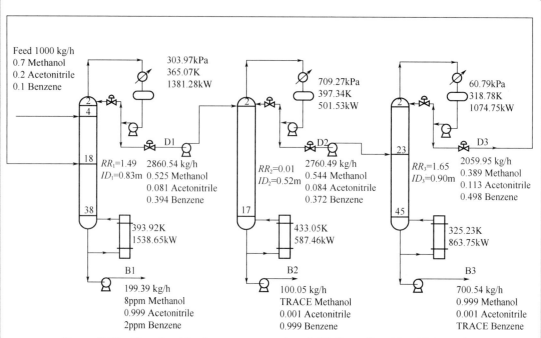

Figure 7.24 Flowsheet for the pressure-swing distillation without heat-integration

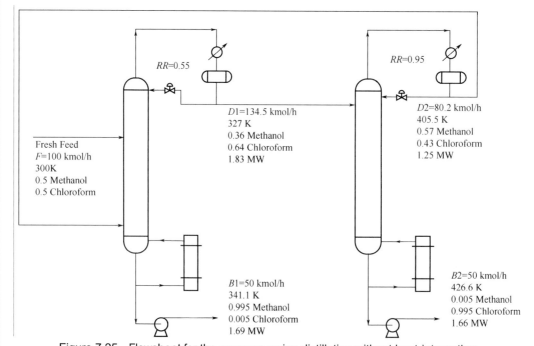

Figure 7.25 Flowsheet for the pressure-swing distillation without heat-integration

References

[1] Sun L Y. Chemical process simulation training—Aspen Plus Course. 2012.

[2] Zhang Z. Process optimization and control strategy of pressure-swing distillation for separating an azeotropic mixture of *n*-heptane and isobutanol[D]. Qingdao: Qingao University of Science and Techenology, 2016: 25-60.

[3] Wang Y, Zhang Z, Xu D, et al. Design and control of pressure-swing distillation for azeotropes with different types of boiling behavior at different pressures[J]. Journal of Process Control, 2016, 42: 59-76.

[4] Zhu Z, Xu D, Liu X, et al. Separation of acetonitrile/methanol/benzene ternary azeotrope via triple column pressure-swing distillation[J]. Separation and Purification Technology, 2016, 169: 66-77.

[5] Luyben W L. Design and control of a fully heat-integrated pressure-swing azeotropic distillation system[J]. Industrial & Engineering Chemistry Research, 2008, 47(8): 2681-2695.

Chapter 8

Ternary Extractive Distillation System Using Mixed Entrainer

8.1 Introduction

In this chapter, the mixture of tetrahydrofuran/ethanol/water containing three binary azeotropes is separated by ternary extractive distillation processes using dimethyl sulfoxide, ethylene glycol and mixed solvent of dimethyl sulfoxide and ethylene glycol as entrainer. Mixed entrainer enhances the flexibility of the tradeoff, because it can only be achieved by changing the composition of the mixed entrainer. The dynamic control of ternary extractive distillation process is complex because of the relatively high number of operating parameters and the interactions between multiple azeotropes.

8.2 Converting from Steady-state to Dynamic Simulation

The extractive distillation process with DMSO is shown in Figure 8.1. The extractive distillation process with mixed entrainer (60% DMSO+40% EG, mole fraction) is shown in Figure 8.2. The ternary ED process requires three distillation columns, in which the first two columns are ED columns (EDCs) and another column is entrainer-recovery column (ERC).

Before exporting to dynamic process from steady-state, the sizes for the major equipment must be specified. According to the commonly used heuristics mentioned by Luyben, the sizes of reflux drum and column sump are set to provide 10 min of liquid holdup when the vessel is full. Adequate pressure drops are given by sizing pumps and valves to satisfy the changes of flow rate.

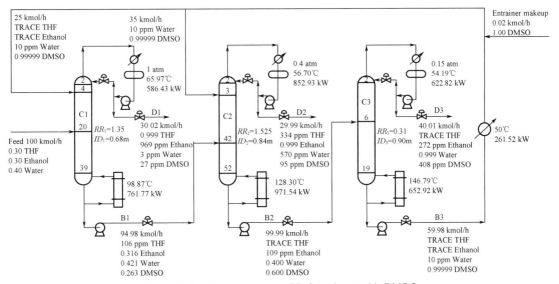

Figure 8.1　Optimal ternary ED flowsheet with DMSO

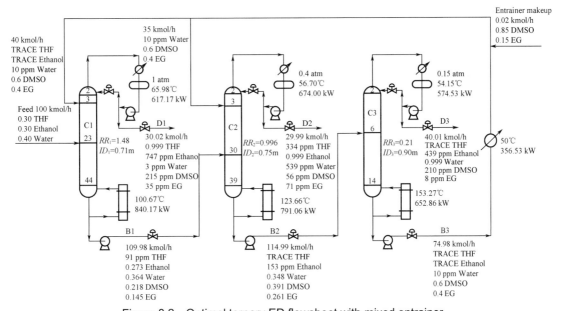

Figure 8.2　Optimal ternary ED flowsheet with mixed entrainer

The slope criterion is used to select temperature-sensitive trays. The temperature and temperature slope profiles for ternary extractive distillation with DMSO are shown in Figure 8.3. In two EDCs, the temperatures at solvent feed stages have a rapid rise and the temperatures at fresh feed stages have a rapid fall. However, these are not good location for temperature control because the fluctuation of feed conditions easily affects the stability of the control system. There are also rapid changes in temperature near the bottom of two EDCs, which are not good location for temperature control because the composition of the key component is not well inferred. It is obvious that stage 15 and 27 also have steep temperature slopes in the EDC1. For the ERC, the slope profile has two peaks at stage 3 and 11. The temperature-sensitive trays in the rectifying section are not suggested to be controlled by reboiler duty due to the lags in response. Thus, the temperatures of stage 27 in the EDC1 and stage 11 in the ERC can be controlled by manipulating corresponding reboiler duties. Because the slope of stage 18 is largest in the temperature difference profiles for the EDC2, it is selected as temperature-sensitive tray.

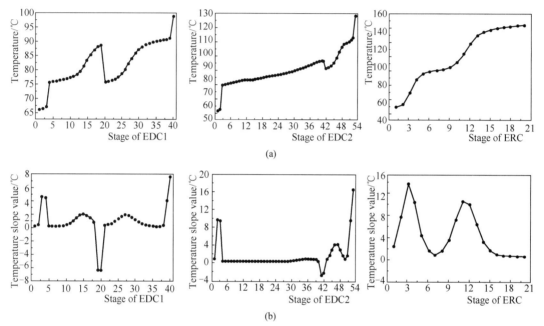

Figure 8.3 The temperature profiles and temperature slope profiles for ternary extractive distillation with DMSO

The temperature and temperature slope profiles for ternary extractive distillation with mixed entrainer are shown in Figure 8.4. It is obvious that stage 17 and 31 have steep temperature slopes in the EDC1. The slope profile of ERC has two peaks at stage 3 and 9. The temperatures of stage 31 in the EDC1 and stage 9 in the ERC can be controlled by manipulating corresponding reboiler duties. For the EDC2, stage 35 is selected as temperature-sensitive tray.

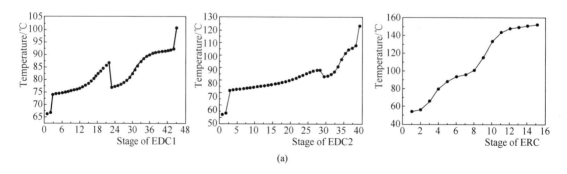

(a)

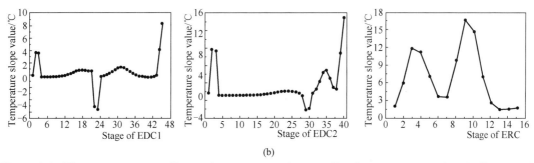

Figure 8.4 The temperature profiles and temperature slope profiles for ternary extractive distillation with mixed entrainer

8.3 Dynamic Control of Ternary Extractive Distillation Process Using Single Solvent

8.3.1 Basic Control Structure

Based on the basic control structure of binary extractive distillation system, the basic control structure of ternary extractive distillation process is proposed. The detailed control loops are listed as follows:

(1) As shown in Figure 8.5, the fresh feed flow rate is controlled to guarantee the constant flow.

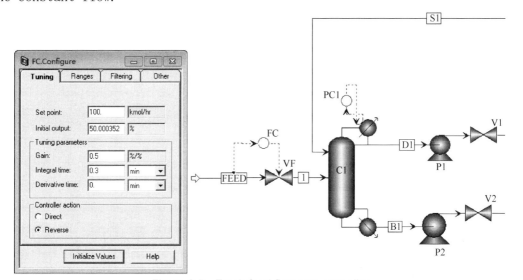

Figure 8.5 Fresh feed flow rate controller

(2) The reflux drum levels are held via the manipulation of distillate flow rate for all columns, as shown in Figure 8.6. The sump levels in the two EDCs are controlled through manipulating the bottom flow rate of two EDCs, as shown in Figure 8.7. The sump level of ERC is controlled via the manipulation of the flow rate of entrainer makeup, as shown in Figure 8.8.

Chapter 8 Ternary Extractive Distillation System Using Mixed Entrainer

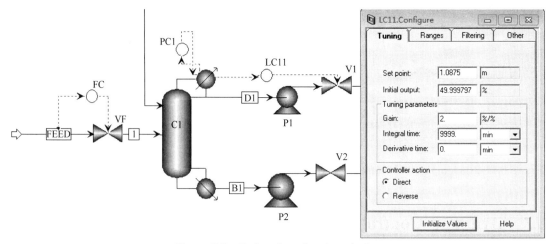

Figure 8.6 Reflux drum level controller

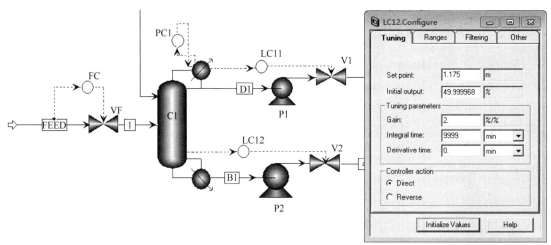

Figure 8.7 Sump levels controller

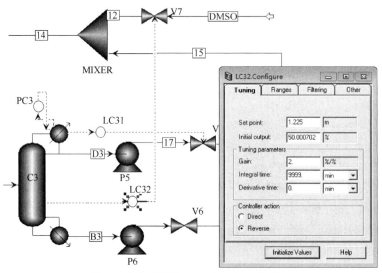

Figure 8.8 ERC sump level controller

(3) The reflux ratios of three columns are fixed, as shown in Figure 8.9 and Figure 8.10.

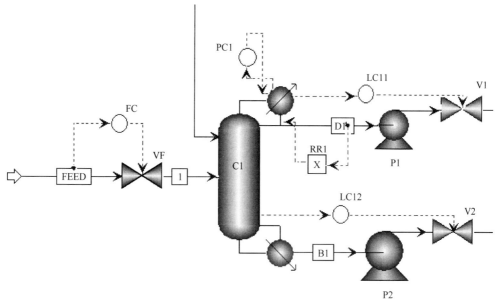

Figure 8.9 Reflux ratio controller

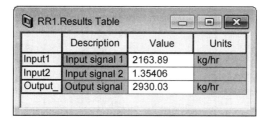

Figure 8.10 The settings of reflux ratio

(4) The proportion of total entrainer flow rate and fresh feed flow rate is constant, as shown in Figure 8.11.

(5) The proportion of the entrainer flow rate to EDC2 and the bottom flow rate of EDC1 are constant.

(6) The temperature of recycled entrainer is maintained at 50 ℃ through manipulating the heat removal rate of cooler.

(7) The temperatures of suitable temperature-sensitive tray in all columns are controlled via the manipulation of the corresponding reboiler duties.

The disturbance rejection capability of proposed control structures is evaluated by introducing ±20% feed flow rate and composition disturbances. All disturbances are added at 0.5 h and the termination time is 20 h. Figure 8.12 shows the dynamic responses of the basic control structure for the ternary extractive distillation with DMSO after introducing ±20% feed flow rate disturbances. The system can arrive at a new steady-state within 3 h and all controlled temperatures can return to the initial value. The purity of three

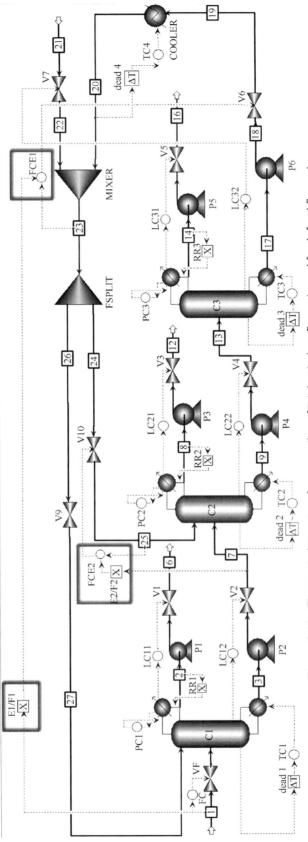

Figure 8.11 The control structure for proportion of total entrainer flow rate and fresh feed flow rate

138 Chemical Process Simulation

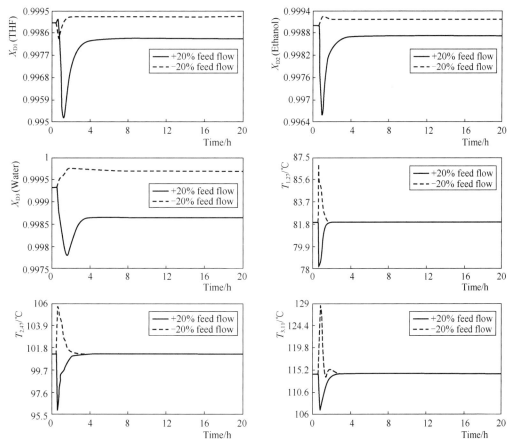

Figure 8.12 The dynamic responses of the basic control structure for the ternary extractive distillation with DMSO after introducing ±20% feed rate disturbances

products are very close to the initial values when encountered −20% feed flow rate disturbance. For +20% feed flow rate disturbance, the purity of THF and water is 99.83% and 99.86% (mole fraction), respectively, having large deviation to their desired values when the new steady-state arrives.

The dynamic responses of the basic control structure for the ternary extractive distillation with DMSO after introducing ±20% feed composition disturbances are shown in Figure 8.13. The composition of THF in mole fraction after introducing +20% feed composition disturbance is 36% THF, 27.4% ethanol, and 36.6% water, while for −20% disturbance the composition is 24% THF, 32.6% ethanol, and 43.4% water. As shown in Figure 8.13, the system is able to arrive at a new steady-state within 3 h after the feed composition disturbances are introduced. The controlled temperatures of two EDCs are brought back to the initial set points, and the temperature of the ERC has a very small fluctuation in a small range. When feed composition disturbance (+20%) is introduced, the product purity of THF and ethanol is maintained quite close to the initial values, while the purity of water has a large deviation from the initial value (99.90% to 99.15%). When the feed composition is changed from 0.3/0.3/0.4 to 0.24/

0.326/0.434 THF/ethanol/water, the purity of THF reduces to 95.24%, which has a large deviation to its desired value 99.9%. Therefore, the basic control structure cannot efficiently handle the disturbances.

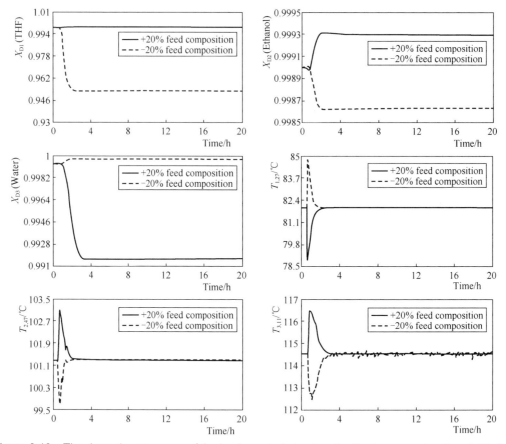

Figure 8.13 The dynamic responses of the basic control structure for the ternary extractive distillation with DMSO after introducing ±20% feed composition disturbances

8.3.2 Dual Temperature Control Structure

It is worth noting that stage 15 in the EDC1 and stage 11 in the ERC are also potential temperature control stages. To obtain better dynamic control performance, an improved dual temperature control structure is explored, and the control structure is shown in Figure 8.14. Two temperature controllers are added to the control structure, in which the temperatures of stage 15 in the EDC1 and stage 3 in the ERC are controlled by corresponding reflux ratio. The other control loops are same with control strategy mentioned in basic control structure.

The dynamic responses of the dual temperature control structure for ternary extractive distillation with DMSO after introducing ±20% feed flow rate disturbances are shown in Figure 8.15. The purity of three products is close to the initial values, and the controlled temperatures are also maintained at

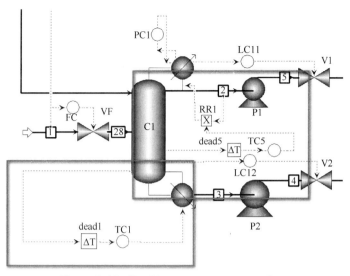

Figure 8.14 Double temperature controllers

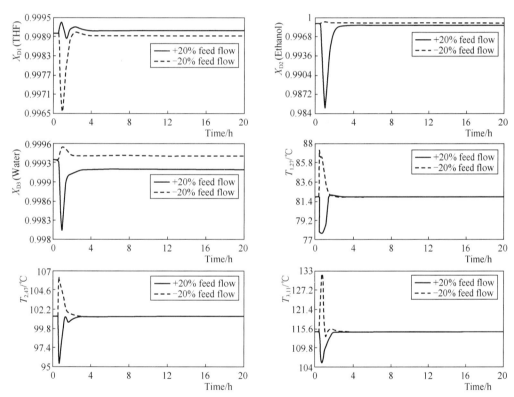

Figure 8.15 The dynamic responses of the dual temperature control structure for the ternary extractive distillation with DMSO after introducing ±20% feed flow rate disturbances

set points. It is obvious that dual temperature control structure has better control performance than basic control structure when faced with same flow rate disturbances. However, all controlled temperatures and the purity of products have large transient deviations from the set points when the feed flow rate disturbances are introduced.

The dynamic responses of the dual temperature control structure for ternary extractive distillation with DMSO after introducing ±20% feed composition disturbances are shown in Figure 8.16. The controlled temperatures of two EDCs are brought back to the set points, and the temperature of the ERC has a very small fluctuation in a small range. After introducing +20% feed composition disturbance, the purity of water is held fairly quite close to the initial value, while the product purity of THF and ethanol has large deviations from the initial values. Therefore, more efficient control structure should be further explored.

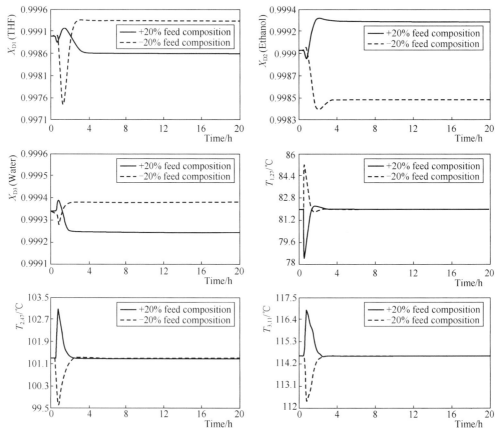

Figure 8.16 The dynamic responses of the dual temperature control structure for the ternary extractive distillation with DMSO after introducing ±20% feed composition disturbances

8.3.3 Composition with Q_R/F Control Structure

The dual temperature control structure is adjusted for some improvements to obtain better dynamic control performance. The composition with the ratio of reboiler duty to mole feed flow rate (Q_R/F) control structure is shown in Figure 8.17. The temperature control loop of stage by reflux ratio in EDC1 is deleted, and the product purity of THF is controlled by reflux ratio. The product purity of ethanol is controlled by reflux ratio in EDC2. To solve the large transient deviations, the feed forward control structure is added, in which the direct control duty of reboiler is replaced by controlling Q_R/F. Composition control loop usually entails larger deadtime with 3 min than temperature control loop.

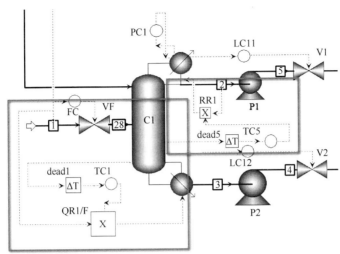

Figure 8.17 The composition with Q_R/F control structure for ternary extractive distillation with DMSO

Figure 8.18 gives the dynamic responses of the composition with Q_R/F control

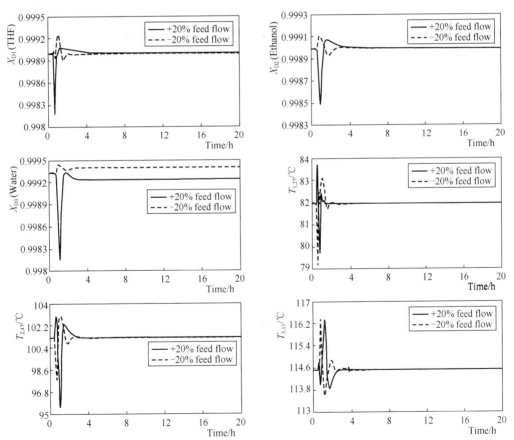

Figure 8.18 The dynamic responses of the composition with Q_R/F control structure for the ternary extractive distillation with DMSO after introducing ±20% feed flow rate disturbances

structure after introducing ±20% feed flow rate disturbances. It is noticed that all product purity is maintained quite close to the initial values after the system arrives at a new steady-state. All product purity has small transient deviations from the initial values. The controlled temperatures of two EDCs are brought back to the initial set points, and the temperature of the ERC has a very small fluctuation in a small range. However, it does not affect the purity of the product. The variations of controlled temperatures are less than 5℃ in EDC1 and EDC2, and 3℃ in ERC. These results of dynamic responses show that the composition with Q_R/F control structure can effectively handle ±20% feed flow rate disturbances.

Figure 8.19 gives the dynamic responses for the composition with Q_R/F control structure after introducing ±20% feed composition disturbances. Three product purity is close to their specified values with small transient deviations from the initial values. The temperature control loop of the ERC has a very small fluctuation in a small range and it does not affect the purity of the product. The variation of controlled temperatures is less than 7℃ in EDC1, 5℃ in EDC2, and 3℃ in ERC. Overall, the composition with Q_R/F control structure shows the good disturbance rejection capability.

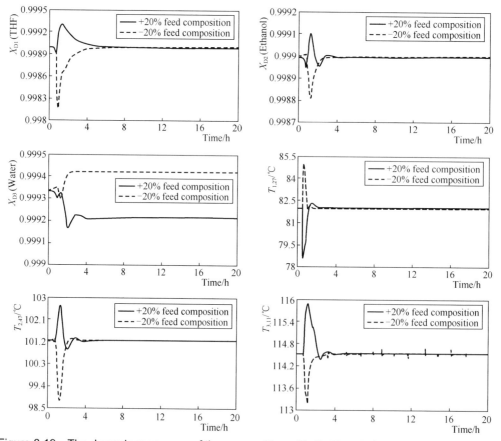

Figure 8.19 The dynamic responses of the composition with Q_R/F control structure for the ternary extractive distillation with DMSO after introducing ±20% feed composition disturbances

8.4 Dynamic Control of Ternary Extractive Distillation Process Using Mixed Entrainer

8.4.1 Basic Control Structure

The proposed basic control structure is also used for ternary extractive distillation process with mixed entrainer. Figure 8.20 shows the dynamic responses of the basic control structure for the ternary extractive distillation with mixed entrainer after introducing ±20% feed flow rate disturbances. The system can arrive at a new steady-state in 3 h, and all controlled temperatures return to the initial values as well. The purity of three products are close to the initial values. However, the product purity of THF and water have large transient deviations relative to the initial values when the feed flow rate increases. The three controlled temperatures have large transient deviation relative to the initial values as well.

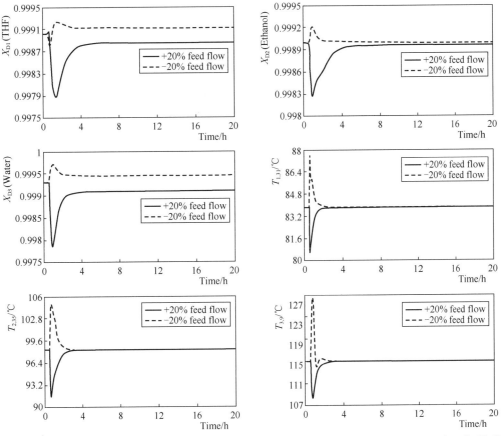

Figure 8.20 The dynamic responses of the basic control structure for the ternary extractive distillation with mixed entrainer after introducing ±20% feed flow rate disturbances

Figure 8.21 shows the dynamic responses of the basic control structure for the ternary extractive distillation with mixed entrainer after introducing ±20% feed composition disturbances. When the feed composition is changed to 0.36/0.274/0.366 of THF/ethanol/water, the product purity of THF and ethanol is close to the specified values, while the purity of water has a large deviation from the initial value (99.93% to 99.65%). When the feed composition is changed to 0.24/0.326/0.434 of THF/ethanol/water, the purity of THF and ethanol is 96.39% and 99.81%, respectively, having large deviation to their specified purity (99.9%) when the new steady-state arrived. Therefore, the basic control structure cannot efficiently handle the disturbances for the ternary extractive distillation with mixed entrainer.

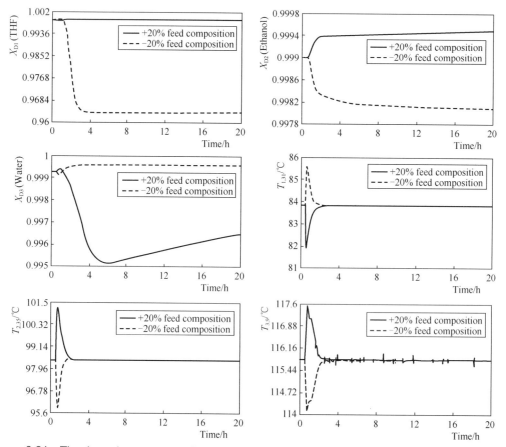

Figure 8.21 The dynamic responses of the basic control structure for the ternary extractive distillation with mixed entrainer after introducing ±20% feed composition disturbances

8.4.2 Composition with Q_R/F Control Structure

To obtain better dynamic control performance for ternary extractive distillation process with mixed entrainer, the improved composition with Q_R/F control structure was also explored. Figure 8.22 gives the dynamic responses for the composition with Q_R/F control structure after introducing ±20% feed flow rate disturbances. All product purity has small transient deviations from the initial values and is maintained quite close to the initial values after the system arrive at a new steady-state. The controlled temperatures are brought back to the initial set points with small deviations. The variations of controlled temperatures are not more than 3℃ in EDC1 and ERC, and the variation is less than 3℃ in EDC2.

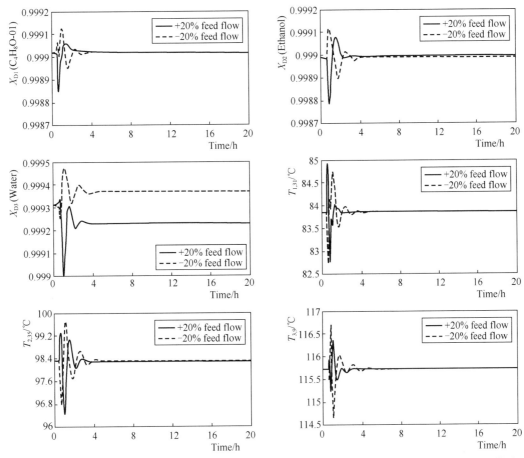

Figure 8.22 The dynamic responses of the basic control structure for the ternary extractive distillation with mixed entrainer after introducing ±20% feed flow rate disturbances

Figure 8.23 gives the dynamic responses for the composition with Q_R/F control structure after introducing ±20% feed composition disturbances. Three product purity is close to their specified values with small transient deviations from the initial values. The controlled temperatures are brought back to the initial set points with small deviations. The temperature control loop of the ERC has a small fluctuation in a small range and it does not affect the purity of the product. The variation of controlled temperatures is not more than 5℃ in EDC1 and EDC2, and less than 3℃ in ERC. Overall, the composition with Q_R/F control structure shows the good disturbance rejection capability.

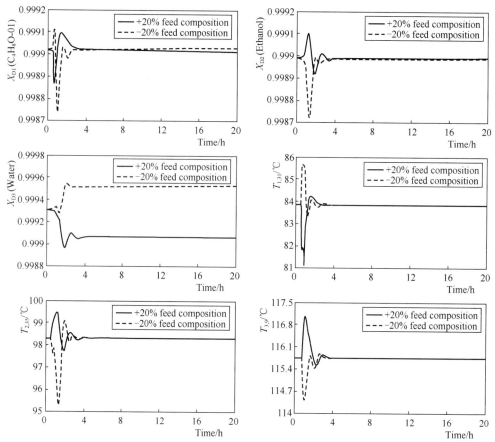

Figure 8.23 The dynamic responses of the basic control structure for the ternary extractive distillation with mixed entrainer after introducing ±20% feed composition disturbances

8.5 Comparisons of the Dynamic Performances of Two Processes

The dynamic performances of ternary extractive distillation process with DMSO and mixed entrainer are compared in this section. Figures 8.24 and 8.25 give the results. It can be observed that the two processes can efficiently handle the disturbances and the purity of products can be held at acceptable values in 3 h after several oscillations.

As shown in Figure 8.24, the extractive distillation process with DMSO has larger transient deviations relative to the initial values in keeping the product purity of THF, ethanol and water when encounters +20% feed flow rate disturbance. There are similar transient deviations in maintaining the product purity of ethanol and water when encounters −20% feed flow rate disturbance, whereas the extractive distillation process with DMSO has larger transient deviation relative to the initial value in keeping the THF product purity.

As shown in Figure 8.25, the extractive distillation process with mixed entrainer has smaller transient deviations relative to the initial values in keeping the product purity of THF after introducing ±20% feed composition

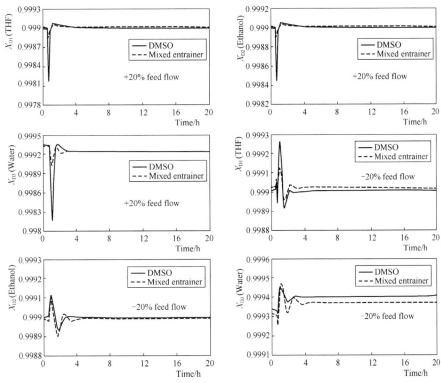

Figure 8.24 Comparisons of dynamic performances between ternary extractive distillation with DMSO and mixed entrainer for feed flow rate disturbances

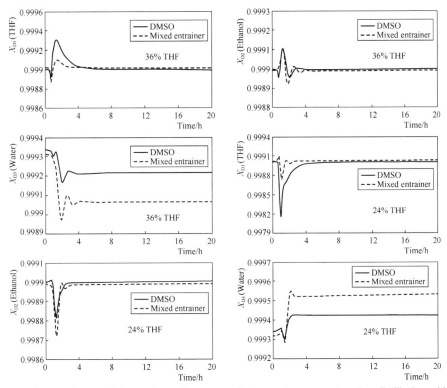

Figure 8.25 Comparisons of dynamic performances between ternary extractive distillation with DMSO and mixed entrainer for feed composition disturbances

disturbances. There are similar transient deviations in maintaining the product purity of ethanol under ±20% feed composition disturbances. However, the extractive distillation process with mixed entrainer has larger transient deviations relative to the initial values in keeping the water product purity under ±20% feed flow rate disturbances. Overall, the ternary extractive distillation process with mixed entrainer shows better dynamic behavior compared with ternary extractive distillation process with DMSO.

Exercises

1. The process of separating benzene and cyclohexane via extractive distillation in a dividing wall column was designed. Furfural was used in the separation of benzene and cyclohexane azeotropic system. Detailed information about the steady-state process is shown in Figure 8.26. Complete the dynamic control of the two processes.

2. The process of separating benzene and acetonitrile was designed, and dimethyl sulfoxide (DMSO) is selected as the entrainer. Detailed information about the steady-state process is shown in Figure 8.27. Complete the dynamic control of the process.

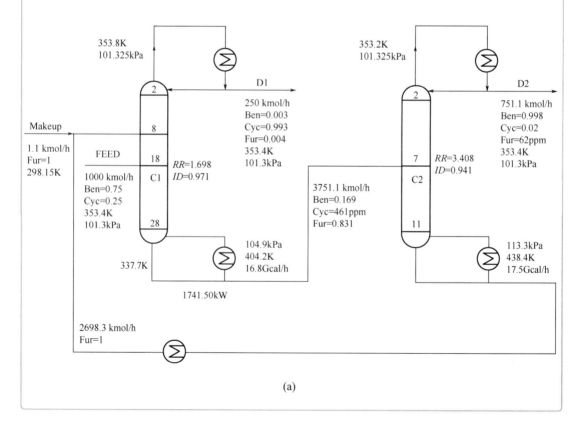

(a)

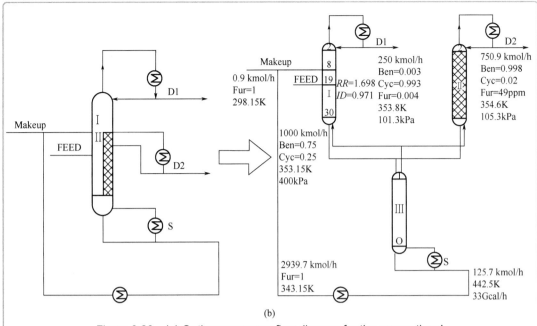

Figure 8.26 (a) Optimum process flow diagram for the conventional configurations; (b) Process flow diagram for the EDWC

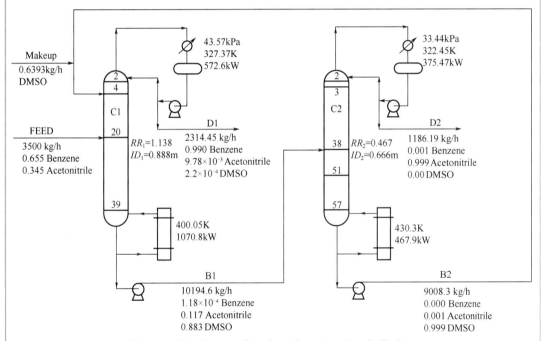

Figure 8.27 Process flowsheet for extractive distillation

3. The process of the separation of the toluene-methanol-water mixture via extractive distillation was designed. Diethylene glycol (DEG) and N-methyl-2-pyrrolidone (NMP) were used as heavy solvent in two processes, respectively. Detailed information about the steady-state process is shown in Figure 8.28 and Figure 8.29. Complete the dynamic control of the two processes.

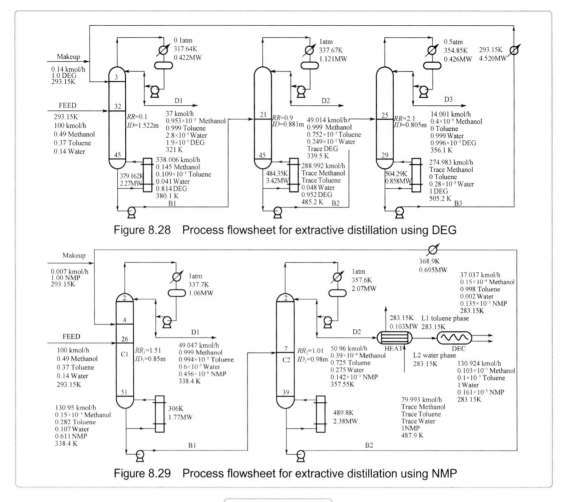

Figure 8.28 Process flowsheet for extractive distillation using DEG

Figure 8.29 Process flowsheet for extractive distillation using NMP

References

[1] Zhang X, Zhao Y, Wang H, et al. Control of a ternary extractive distillation process with recycle splitting using a mixed entrainer[J]. Industrial & Engineering Chemistry Research, 2017, 57(1): 339-351.

[2] Zhao Y, Zhao T, Jia H, et al. Optimization of the composition of mixed entrainer for economic extractive distillation process in view of the separation of tetrahydrofuran/ethanol/water ternary azeotrope[J]. Journal of Chemical Technology & Biotechnology, 2017, 92(9): 2433-2444.

[3] Luyben W L. Distillation design and control using Aspen simulation[M]. John Wiley & Sons, 2013.

[4] Luyben W L. Evaluation of criteria for selecting temperature control trays in distillation columns[J]. Journal of Process Control, 2006, 16(2): 115-134.

[5] Luyben W L. Plantwide control of an isopropyl alcohol dehydration process[J]. AIChE Journal, 2006, 52(6): 2290-2296.

[6] Sun L, Wang Q, Li L, et al. Design and control of extractive dividing wall column for separating benzene/cyclohexane mixtures[J]. Industrial & Engineering Chemistry Research, 2014, 53(19): 8120-8131.

[7] Yang S, Wang Y, Bai G, et al. Design and control of an extractive distillation system for benzene/acetonitrile separation using dimethyl sulfoxide as an entrainer[J]. Industrial & Engineering Chemistry Research, 2013, 52(36): 13102-13112.

[8] Wang Y, Zhang X, Liu X, et al. Control of extractive distillation process for separating heterogeneous ternary azeotropic mixture via adjusting the solvent content[J]. Separation and Purification Technology, 2018, 191: 8-26.

Chapter 9

Hybrid Process Including Extraction and Distillation

9.1 Introduction

Liquid-liquid extraction, or simply solvent extraction, is a method to separate compounds based on their relative solubilities in two different immiscible liquids, which are usually water and an organic solvent. It is an extraction of a substance from one liquid into another liquid phase. Liquid-liquid extraction is a basic technique in chemical industries. Solvent extraction is used in roasted coffee beans, ore processing, the production of fine organic compounds, the processing of perfumes, and the production of vegetable oils and food flavoring agents.

The dehydration of PM via traditional heterogeneous azeotropic distillation and extractive distillation processes were investigated by Chen et al. Although these processes successfully achieved the separation of PM and water, the distillation column required to achieve the pre-separation consuming excessive energy. To save energy, the dehydration of PM via hybrid extractive distillation was investigated in this chapter. Based on the principle of selectivity and distribution coefficient, chloroform and 2-EA were selected as solvents. In order to save more energy, a mixture of solvent (chloroform, 2-EA) was used to separate. In addition, the corresponding dynamic control was studied.

9.2 Solvent Selection

For the PM recovery system, liquid-liquid extraction combined with either heterogeneous azeotropic distillation or extractive distillation processes are designed. The principal advantage of both processes is that PM and water can be pre-separated through liquid-liquid extraction without energy consumption. Thus, the selection of an effective solvent for extraction is one of the most important parts in the design of the two hybrid processes. Normally, the distribution coefficient (i.e., the ratio of concentrations of PM in the extract and raffinate equilibrium) is a vital performance index for the solvent in liquid-liquid extraction. The solvent should have stronger affinity for PM than water, so that PM can be extracted from the fresh feed during liquid-liquid extraction. Therefore, the distribution coefficient must be greater than 1.

Based on the above criteria, 2-ethylhexanoic acid (2-EA), n-decyl alcohol, isopropyl acetate, chloroform, ethyl acetate, and vinyl acetate is screened. For all the studies, simulations are done using Aspen Plus. After introducing the fresh feed and solvent candidates to the extractor, simulations were run in Aspen Plus. The solvent-to-feed mole ratios (0.5, 0.75, 1, and 1.25) are varied while the flow rate of the fresh feed is held constant for all processes. Figure 9.1 illustrates the PM's equilibrium curves for the four mole ratios. Among the six solvents, 2-EA exhibits favorable extraction performance with very low PM concentrations in the aqueous phase at each mole ratio, while chloroform exhibits favorable extraction performance with very high PM concentrations in the solvent-rich phase at each mole ratio.

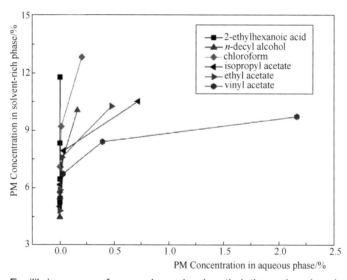

Figure 9.1 Equilibrium curves for propylene glycol methylether using six solvents at 45℃

For all the studies, the non-random two-liquid (NRTL) thermodynamic model are selected to describe the phase behavior of the PM-water-solvent system. Concurrently, the UNIFAC group-contribution method is used to estimate the missing parameters for the PM and solvents. Additionally, the nonideality of the system should be considered and the accuracy of the model should be validated. Due to the importance of the liquid-liquid extraction for the processes of liquid-liquid extraction combined with either heterogeneous azeotropic distillation or extractive distillation, liquid-liquid equilibrium experiments for the two ternary systems (water-PM-chloroform and water-PM-2-ethylhexanoic acid) are performed. All NRTL parameters are listed in Table 9.1.

Table 9.1 NRTL model parameters of PM-water-solvent systems

Comp. i	PM	PM	water	PM	water
Comp. j	water	chloroform	chloroform	2-EA	2-EA
a_{ij}	-1.530	0	8.844	0	0
a_{ji}	7.626	0	-7.352	0	0
b_{ij}(K)	431.045	-474.553	-1140.115	-71.074	2284.685
b_{ji}(K)	-1998.746	891.344	3240.688	79.497	312.536
c_{ij}	0.300	0.300	0.200	0.300	0.200
Comp. i	PM	water	PM	water	PM
Comp. j	n-decyl alcohol	n-decyl alcohol	isopropyl acetate	isopropyl acetate	ethyl acetate
a_{ij}	0	-5.577	0	26.900	0
a_{ji}	0	-7.560	0	-1.423	0
b_{ij}(K)	391.239	5080.560	-60.437	-6530.301	-125.696
b_{ji}(K)	-82.605	2510.615	338.092	618.519	374.4926
c_{ij}	0.300	0.200	0.300	0.200	0.300
Comp. i	water	PM	water		
Comp. j	ethyl acetate	vinyl acetate	vinyl acetate		
a_{ij}	9.463	0	0		
a_{ji}	-3.710	0	0		
b_{ij}(K)	-1705.68	36.687	1364.600		
b_{ji}(K)	1286.138	249.948	415.700		
c_{ij}	0.200	0.300	0.200		

9.3 Simulation of the Extraction Combined with Distillation Process

9.3.1 Extraction Combined with Heterogeneous Azeotropic Distillation Process (LEHAD)

Figure 9.2 shows the conceptual liquid-liquid extraction combined with heterogeneous azeotropic distillation process. Figure 9.3~Figure 9.6 show the feed flow information and distillation column parameters. The fresh feed (F), comprising 92.2% (mole fraction) water and 7.8% PM at a rate of 1000 kmol/h, is the same as that used by Chen et al. It is sent into a liquid-liquid extraction column (LEC1). Because chloroform has a much greater density than F, it is sent into the LEC1 at the first stage and the F is sent into the LEC1 column at the last stage. Then, the PM is extracted from the fresh feed and the extract phase (E) is sent to a stripper (C2) to obtain high-purity PM as the bottom stream.

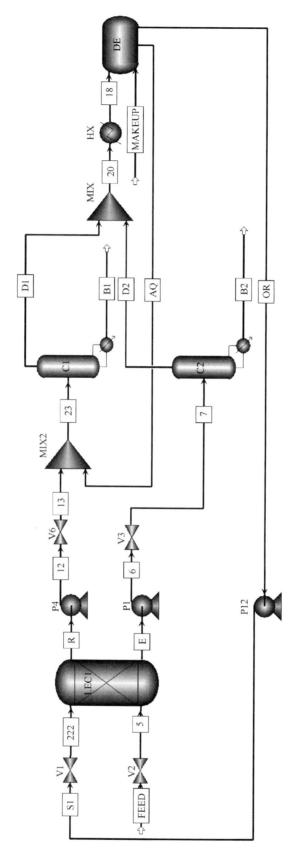

Figure 9.2 Conceptual design flowsheet for the liquid-liquid extraction combined with heterogeneous azeotropic distillation process

156 Chemical Process Simulation

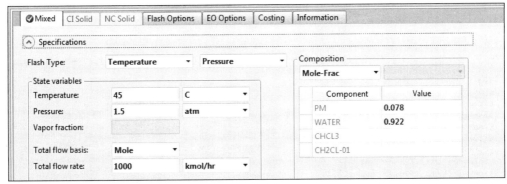

Figure 9.3　Inputting the information of feed stream

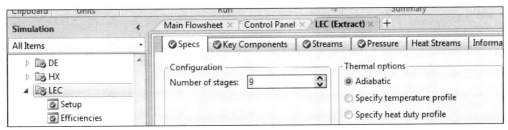

Figure 9.4　Parameters of liquid-liquid extraction column

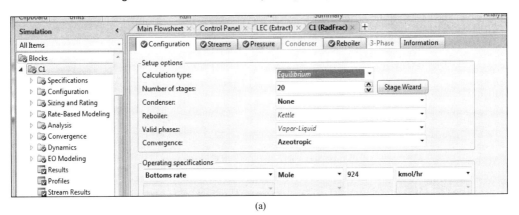

(a)

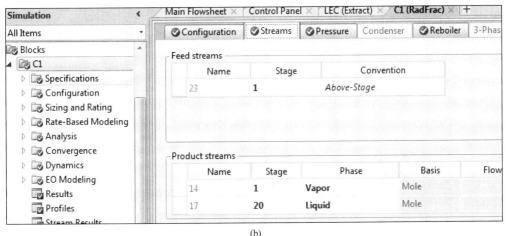

(b)

Figure 9.5　Parameters of C1 column

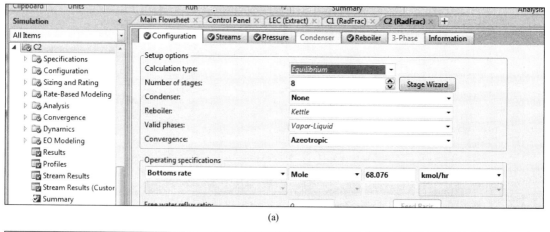

(a)

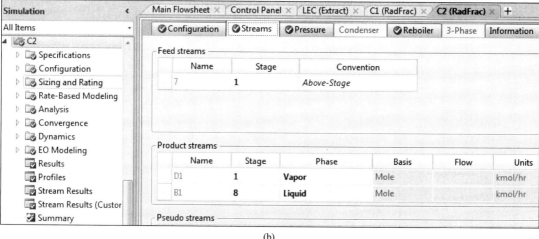

(b)

Figure 9.6   Parameters of C2 column

The raffinate phase (R), which contains mostly water, is sent into another stripper (C1) to achieve further separation of the water as the bottom stream. The top vapor streams from C1 and C2 (D2) are mixed into one stream by a mixer, cooled by cooling water to 45℃, and fed into a decanter where natural liquid-liquid separation proceeds. The organic phase (OR) of almost pure chloroform is recycled to the LEC1. The aqueous phase (AQ) consists mostly of water and it is sent into the LEC1 as a feed stream after mixing with R obtained from C1. Figure 9.7~Figure 9.9 show the simulation results where output streams are shown.

(1) Input the information of feed stream.

(2) There are 9 stages in the liquid-liquid extraction column, and the thermal options is adiabatic.

(3) Input the parameters of C1 column.

(4) Input the parameters of C2 column.

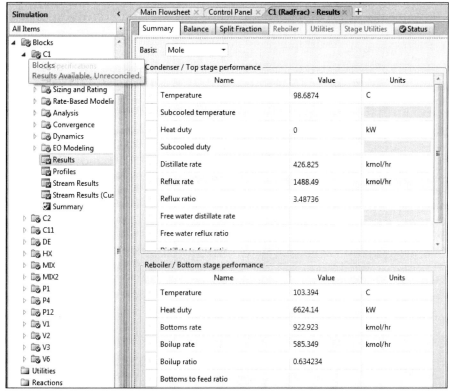

Figure 9.7　Simulation results of C1 column

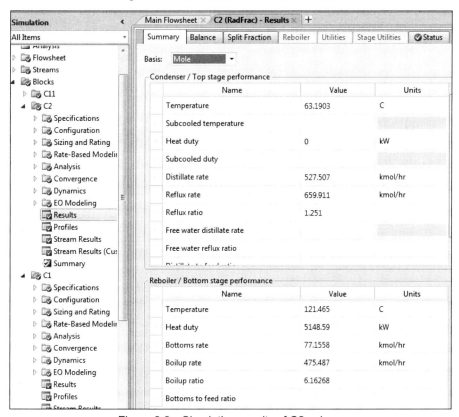

Figure 9.8　Simulation results of C2 column

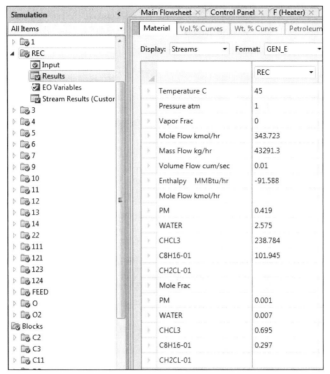

Figure 9.9   Simulation results of REC stream

9.3.2   Extraction Combined with Extractive Distillation Process (LEED)

The conceptual liquid-liquid extraction combined with extractive distillation process is shown in Figure 9.10. The flowsheet includes a liquid-liquid extraction column (LEC2), a extractive distillation column (EDC), and a recovery column (RC). The LEC2 is equivalent to a pre-separation column, the fresh feed (F, which is the same as in the previous process) is sent into the LEC2 at the first stage and 2-EA (S1) is sent into the LEC2 at the last stage. From the RCM of PM-water-2-EA, nearly pure water can be obtained in the raffinate phase (R), and a ternary mixture of 2-EA-PM-water is obtained in the extract phase (E). Figure 9.11 and Figure 9.12 show the distillation column parameters. In this flowsheet, the raffinate phase with a water purity of 99.9% is outputted directly and E is sent into the EDC. For the extractive distillation, 2-EA is selected as the solvent to achieve separation by altering the relative volatilities of the PM and water. By adding moderate 2-EA (S2) at the upper part of the EDC, water as the top product of the EDC (D1) is obtained with a purity of 99.8%, and the mixture of PM and 2-EA as the bottom product (B1) is sent into the RC to achieve further separation. In the RC, PM is obtained as the top product (D2) with a purity of 99.9% and almost pure 2-EA is obtained as the bottom product (B2). B2 is cooled to 45℃ by cooling water and mixed with supplemental 2-EA.

液液萃取-萃取精馏
分离丙二醇甲基醚
和水的稳态工艺

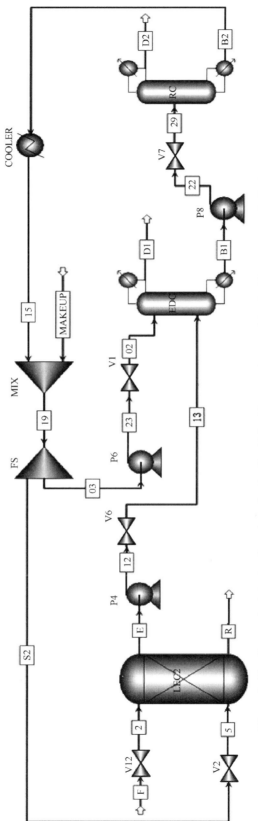

Figure 9.10 Conceptual design flowsheet for the liquid-liquid extraction combined with extractive distillation process.

Chapter 9 Hybrid Process Including Extraction and Distillation 161

Then 2-EA is split into two streams with specific molar weights for recycling to the LEC2 and EDC, respectively. Figure 9.13 and Figure 9.14 show the simulation results where output streams are shown.

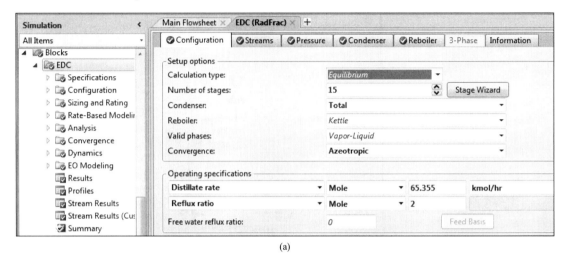

(a)

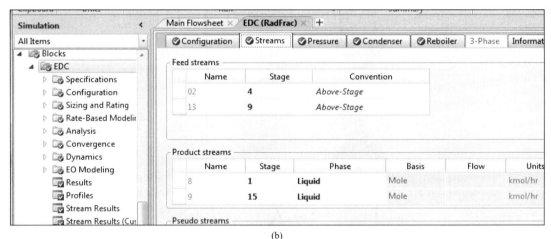

(b)

Figure 9.11   Parameters of EDC column

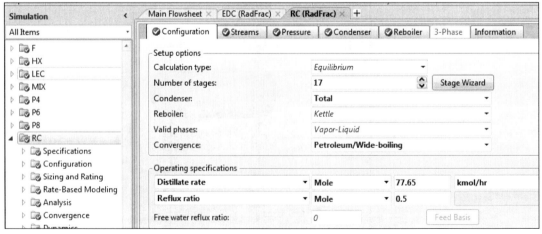

(a)

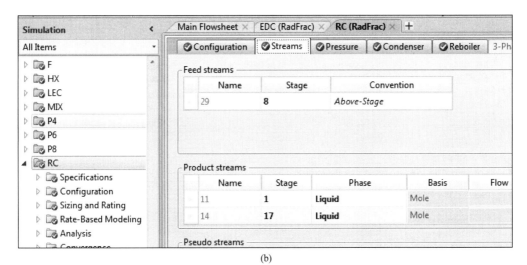

(b)

Figure 9.12　Parameters of RC column

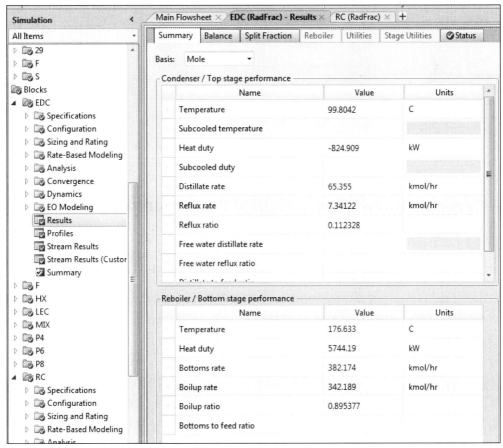

Figure 9.13　Simulation results of EDC column

(1) The number of stages is 15 in the liquid-liquid extraction column.
(2) Inputting the parameters of EDC column.
(3) Inputting the parameters of RC column.

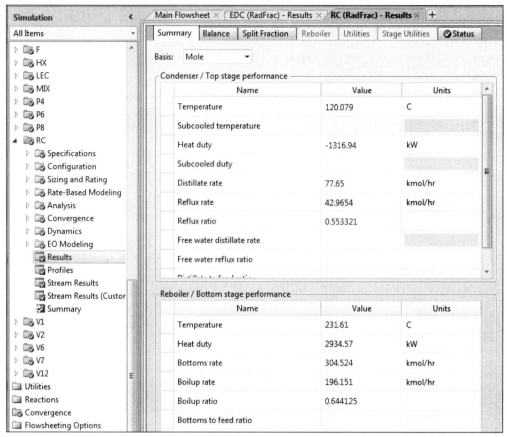

Figure 9.14　Simulation results of RC column

9.4　Dynamic Simulation of Hybrid Extraction-distillation

9.4.1　Selection of Temperature-sensitive Trays

The detailed process is shown in Figure 9.15 and Figure 9.16. For dynamic control, selecting the temperature-sensitive tray is crucial. The slope criterion is the most common method. However, the slope criterion does not consider the interaction with the possible manipulated variables available for the control loops. Therefore, three methods are used to select the temperature-sensitive tray for the two processes. For the LEHAD process, the slope criterion and open-loop sensitivity criterion methods are used to determine the tray temperature control point for C1 and C2 column. For open-loop sensitivity, the tray with a larger temperature difference should be selected. The slope criterion and open-loop sensitivity test results are shown in Figure 9.17. Stage 14 of the C1 column and stage 4 of the C2 column can be selected for temperature control. For the LEED process, the slope criterion and SVD methods are used to select the temperature-

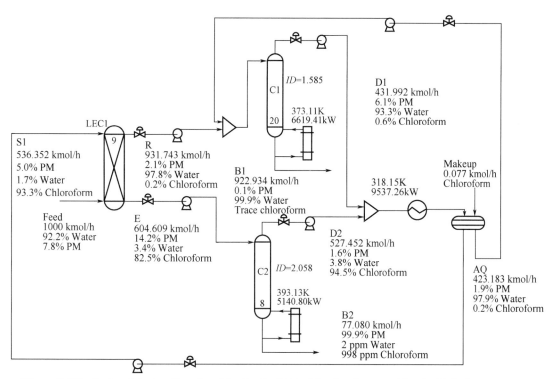

Figure 9.15　Optimal design flowsheet for the liquid-liquid extraction combined with heterogeneous azeotropic distillation process(LEHAD)

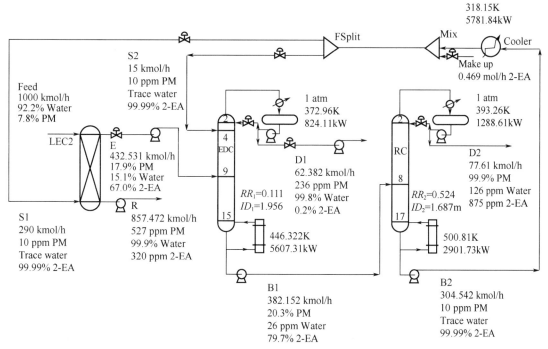

Figure 9.16　Optimal design flowsheet for the liquid-liquid extraction combined with extractive distillation process(LEED)

Chapter 9　Hybrid Process Including Extraction and Distillation

sensitive tray. As shown in Figure 9.18, for SVD methods, the dot line is associated with reflux, and the square line is associated with heat input. The results show that stage 12 can be controlled by heat input in the EDC column. For the RC column, stage 3 can be controlled by reflux and stage 11 by heat input. The results using SVD are similar to the slope criterion results. First, stage 11 is determined to be the temperature-sensitive tray by adjusting the reboiler duties in the RC. The information of converting is from steady-state to dynamic simulation.

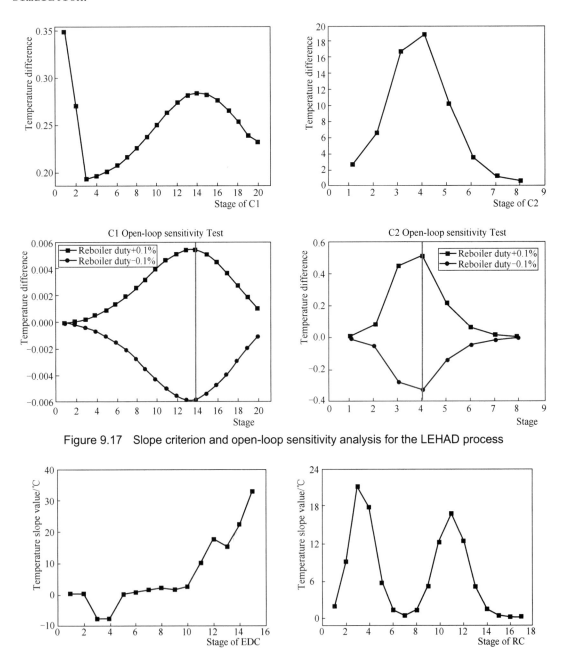

Figure 9.17 Slope criterion and open-loop sensitivity analysis for the LEHAD process

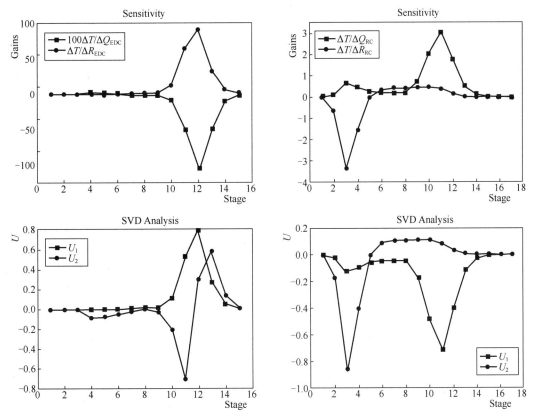

Figure 9.18 Slope criterion and singular value decomposition analysis for the LEED process

9.4.2 Dynamic Control of the LEHAD Process

9.4.2.1 Basic Control Structure of the LEHAD Process

Based on the steady-state flowsheet, a basic control strategy is proposed for the LEHAD process. The basic controllers are used, and the related settings for the basic control are as follows:

(1) The fresh feed flow rate is controlled via flow controllers.

(2) The solvent flow rate is controlled via flow controllers, and there is a constant ratio with the fresh feed flow rate. Meanwhile, the flow controller is in cascade mode.

(3) The operating pressure in the two strippers is controlled by operating the top vapor flow rate.

(4) The sump levels in the two strippers are controlled by operating the bottom flow rate.

(5) The organic phase level of the decanter is controlled by operating the solvent makeup flow rate.

(6) The aqueous phase level of the decanter is controlled by operating the aqueous phase stream outlet flow rate.

(7) The temperatures of stage 14 in C1 and stage 4 in C2 are controlled by operating the corresponding reboiler duty.

(8) The temperature of the mixed stream leaving the cooler is controlled by operating the corresponding cooler heat removal.

Proportional-integral (PI) controllers are used for the flow, pressure, and temperature controllers. For the level loops in the two strippers, proportional-only (P-only) controllers are used, and the gain (K_C) and integral time (τ_I) are set as 2 and 9999 min, respectively. For the decanter organic level control loops, the gain is set to 10 to ensure the adjustment of the import and export of the organic phase without delay because the level is controlled via the small solvent makeup flow rate. For the flow controllers and pressure controllers, K_C is 0.5 and 20, τ_I is 0.3 and 12, respectively. Each temperature controller is inserted with a 1 min deadtime block. In this work, relay-feedback tests are conducted, and the closed loop is selected as the test method for all temperature controllers to obtain the ultimate gain (K_U) and period (P_U). Then, the corresponding tuning parameters for the temperature controllers are calculated using the Tyreus-Luybe rule, as shown in Table 9.2. Figure 9.19 shows the basic control strategy of the LEHAD process.

Table 9.2 Tuning parameters of basic control structure for LEHAD process

Controller	TC1	TC2	TC3
Controller action	Reverse	Reverse	Reverse
Controlled variable	$T_{1,14}$	$T_{2,4}$	$T_{stream18}$
Manipulated variable	Q_{C1}	Q_{C2}	Q_{cooler}
Transmitter range/°C	0~203.63	0~197.85	0~90.04
Controller output range / (GJ/h)	0~47.32	0~0.037	−68.21~0
K_U	38.37	1.12	0.224
P_U / min	2.4	4.20	1.8
K_C	11.99	0.35	0.07
τ_I / min	5.28	9.24	3.96

The ±20% feed flow rate and ±20% feed composition disturbances are added to evaluate the control performance of the basic control strategy of the LEHAD process. All disturbances are set to be introduced at 0.5 h and finished at 20 h. The corresponding dynamic responses for the LEHAD process in the case of adding ±20% feed rate disturbances are shown in Figure 9.20. The purity of PM is stabilized at 99.9%, which is almost the initial specification, and water purity is stabilized at 99.8%, which deviates from the initial value. All controlled temperatures of C1 and C2 can also stabilize at the initial value.

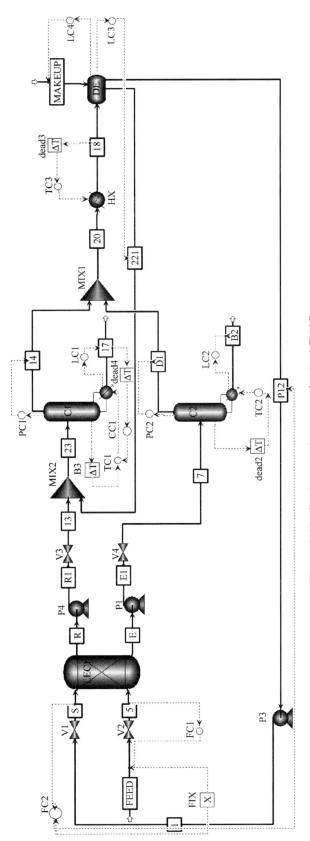

Figure 9.19 Basic control structure for the LEHAD process

Chapter 9 Hybrid Process Including Extraction and Distillation

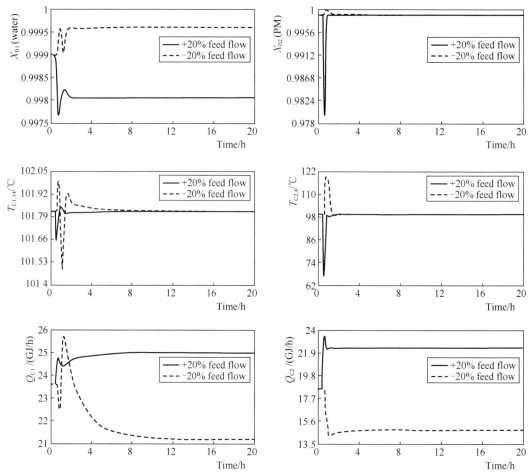

Figure 9.20 Dynamic responses of the basic control structure for the LEHAD process after introducing ±20% feed flow rate disturbances

The controlled temperature in C2 has a transient deviation of more than 30 ℃ after fresh feed flow rate disturbances are added, which leads to a large transient deviation in PM purity. This is due to a lag in the response signal of the temperature controller when the temperature changes in stage 4 in C2. Therefore, the reboiler duty lags behind the variation in tray temperature.

Figure 9.21 shows the dynamic responses for the LEHAD process after introducing ±20% feed composition disturbances. For the +20% disturbance, the fresh feed contains 9.36% PM and 90.64% water, whereas that for the −20% disturbance consists of 6.24% PM and 93.76% water. As shown in Figure 9.21, the PM and water purity can be near the initial values within 3 h after introducing the composition disturbances and maintain a steady-state. The controlled temperatures in the two strippers are stable at the initial set points. Overall, a corresponding improved control strategy should be proposed to maintain the water purity and eliminate the lag effect.

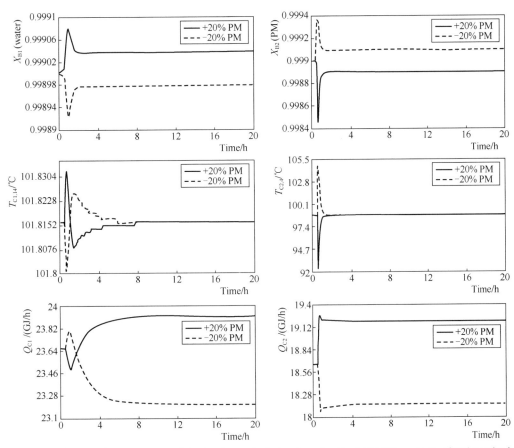

Figure 9.21 Dynamic responses of the basic control structure for the LEHAD process after introducing ±20% feed composition disturbances

9.4.2.2 Improved Control Structure of the LEHAD Process

An improved control structure for the LEHAD process is explored. The composition–temperature cascade control structure is used to maintain the water purity of stream B1 when introducing ±20% feed flow rate disturbances. The dead time for the composition controller is usually larger than the temperature controller, which is set as 3 min. The input signal is the water purity of stream B1, and it is in cascade with the temperature controller to ensure the water purity and quick response. To reduce the lag time, the feed forward control structure is added. The reboiler duty in C2 and the feed flow rate are proportional to each other, and a multiplier which represents the ratio of reboiler duty to mole feed flow rate (Q_{RC2}/F) is added. For the Q_{RC2}/F control structure, one input signal is the ratio that is controlled via the temperature controller of stage 4 in C2, and the other is the fresh feed mole flow rate. After the relay-feedback test, the ultimate K_U and P_U of the two temperature controllers for the two strippers and the composition controller are calculated via Tyreus–Luyben tuning. The tuning parameters are shown in Table 9.3. Figure 9.22 shows the final control strategy

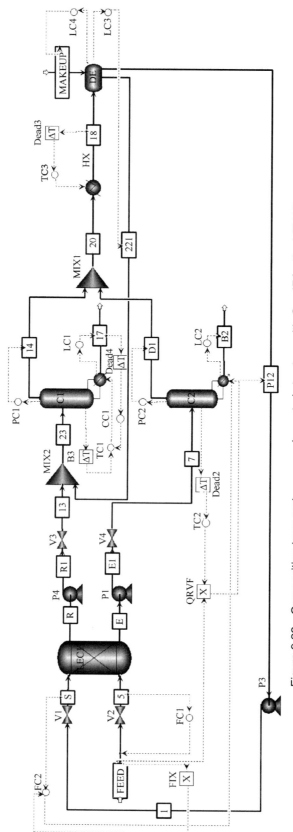

Figure 9.22 Composition-temperature cascade control structure with Q_{RC2}/F for the LEHAD process

for the LEHAD process, and the dynamic responses are shown in Figure 9.23 and Figure 9.24. The water purity stabilized at approximately 99.9% for the ±20% feed flow rate disturbances. The transient deviation of stage 4 temperature in C2 is below 12℃ when fresh feed flow rate disturbances are introduced. Thus, the composition-temperature cascade control structure with Q_{RC2}/F shows good controllability for corresponding feed disturbances.

Table 9.3 Tuning parameters of improved control structure for LEHAD process

Controller	TC1	TC2	TC3	CC1
Controller action	Reverse	Reverse	Reverse	Reverse
Controlled variable	$T_{1,14}$	$T_{2,4}$	$T_{stream18}$	$X(water)_{1,B1}$
Manipulated variable	Q_{C1}	Q_{C2}	Q_{cooler}	$T_{1,14}$
Transmitter range / ℃	0~203.63	0~197.85	0~90.04	0~1.998
Controller output range / (GJ/h)	0~47.32	0~0.037	-68.21~0	0~203.63
K_U	40.02	1.12	0.224	57.18
P_U / min	2.4	4.2	1.8	16.2
K_C	13.13	0.35	0.07	17.87
τ_I / min	5.28	9.24	3.96	35.64

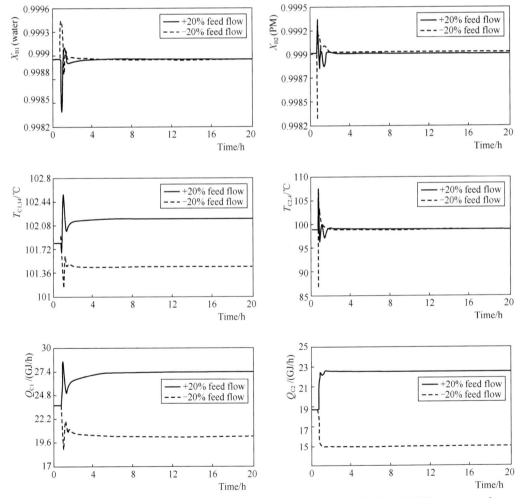

Figure 9.23 Dynamic responses of the improved control structure for the LEHAD process after introducing ±20% feed flow rate disturbances

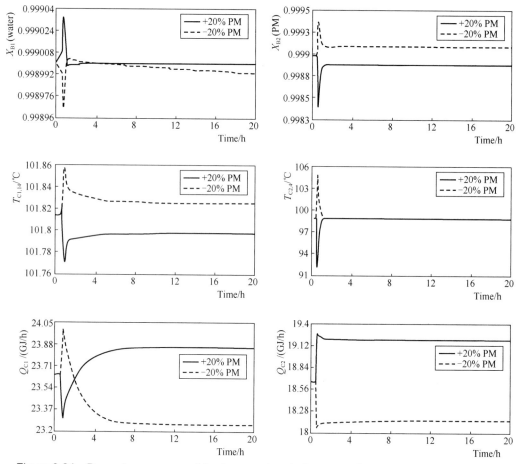

Figure 9.24 Dynamic responses of the improved control structure for the LEHAD process after introducing ±20% feed composition disturbances

9.4.3 Dynamic Control of the LEED Process

9.4.3.1 Basic Control Structure of the LEED Process

For the LEED process, the initial control structure is shown in Figure 9.25. The settings for some controllers are different from the LEHAD process, and the detailed differences are as follows:

(1) The operating pressures in the ERC and RC are controlled by operating the heat removal rate of the corresponding condensers.

(2) The reflux drum levels in the ERC and RC are controlled by operating the flow rate of the distillates.

(3) The sump level in the RC is controlled by operating the flow rate of the solvent makeup.

(4) The reflux ratio in the ERC and RC is fixed.

(5) The temperatures of stage 12 in the EDC and stage 11 in the RC are controlled by operating the reboiler duty.

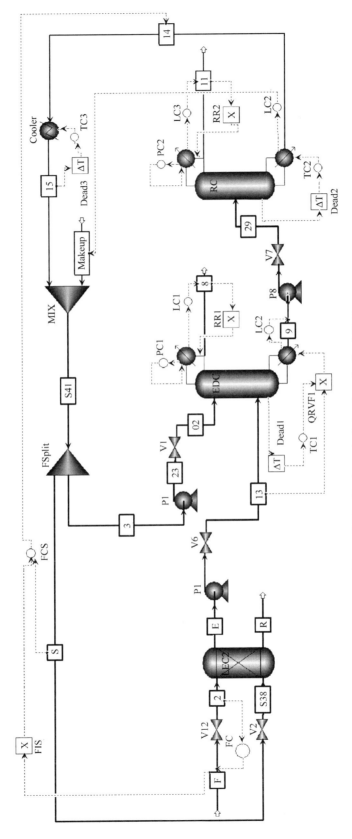

Figure 9.25 Basic control structure for the LEED process

(6) The temperature of the recycled solvent is controlled by operating the heat removal rate of the cooler.

For the three temperature controllers in the ERC and RC, the tuning parameters calculated using the Tyreus-Luybe rule are listed in Table 9.4. The corresponding dynamic responses for the LEED process from adding the ±20% feed flow rate and ±20% composition disturbances are shown in Figure 9.26 and Figure 9.27, respectively. The water purity in the bottom stream of LED2 (R) and the top stream of EDC (D1) can approach the initial purity when achieving a new stability.

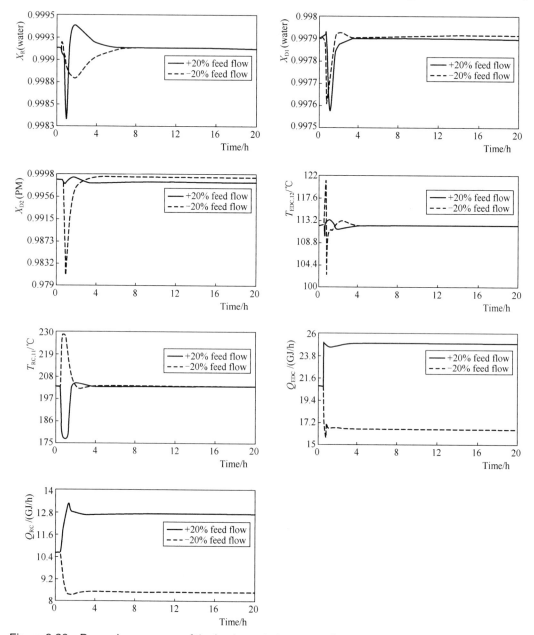

Figure 9.26 Dynamic responses of the basic control structure for the LEED process after introducing ±20% feed flow rate disturbances

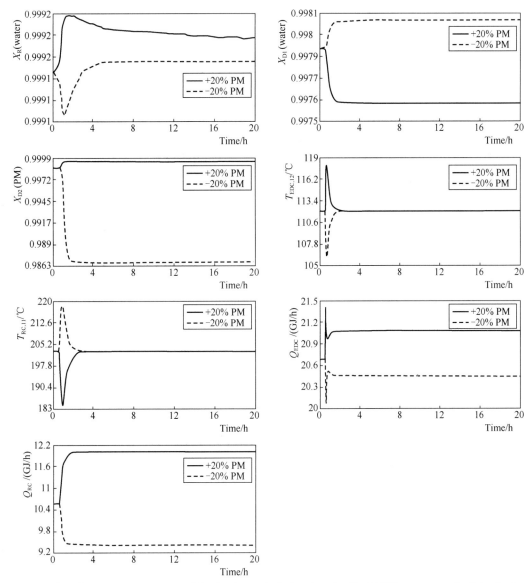

Figure 9.27 Dynamic responses of the basic control structure for the LEED process after introducing ±20% feed composition disturbances

However, the PM purity is 98.63%, which has a large deviation from the initial purity when a −20% feed composition disturbance is encountered. Moreover, the transient deviation for the controlled temperature of stage 11 in the RC is more than 20 ℃ when fresh feed disturbances are introduced. Therefore, the basic control structure cannot handle the disturbances efficiently, and a corresponding improved control structure should be studied.

Table 9.4 Tuning parameters of basic control structure for LEED process

Controller	TC1	TC2	TC3
Controller action	Reverse	Reverse	Reverse
Controlled variable	$T_{1,12}$	$T_{2,11}$	$T_{stream14}$
Manipulated variable	Q_{EDC}/F	Q_{RC}	Q_{cooler}

Chapter 9 Hybrid Process Including Extraction and Distillation

Continued

Controller	TC1	TC2	TC3
Transmitter range / ℃	0~223.92	0~405.33	0~89.87
Controller output range / (GJ/h)	0~0.096	0~21.13	-42.68~0
K_U	1.6	1.82	0.26
P_U / min	6.48	7.20	2.4
K_C	0.50	0.57	0.08
τ_I / min	14.52	15.84	5.28

9.4.3.2 Improved Control Structure of the LEED Process

In the RC, it is not easy to achieve light and heavy component purity control in the top and bottom parts using only one control tray due to the large temperature difference. Thus, the temperature of stage 3 is controlled via the reflux ratio to prevent 2-ethylhexanoic acid from rising to the top. This is because there is a lag in the response of the temperature controller when the temperature changes in stage 11. Therefore, the reboiler duty lags behind the variation in tray temperature. Thus, the ratio of reboiler duty to mole feed flow rate (Q_{RRC}/F) can reduce the lag time. The two temperature controllers in the RC are tuned individually. For these temperature controllers, the temperature controller for stage 11 is tuned primarily, and the temperature controller for stage 3 is tuned afterward. For the four temperature controllers in the ERC and RC, the tuning parameters are listed in Table 9.5. The improved dual temperature control structure with Q_{RRC}/F is shown in Figure 9.28 and is evaluated via the feed flow rate and the feed composition disturbances. Figure 9.29 and Figure 9.30 provide the dynamic responses of the improved control structure with dual temperature and Q_{RRC}/F for the LEED process by adding ±20% feed disturbances. The PM product composition is maintained at 99.84%, which is near the initial values. The fluctuations of controlled temperature in the RC are greatly improved compared to the basic control structure performance when fresh feed disturbances are introduced. Hence, the improved dual temperature control structure with Q_{RRC}/F can handle the disturbances well.

Table 9.5 Tuning parameters of improved control structure for LEED process

Controller	TC1	TC2	TC3	TC4
Controller action	Reverse	Reverse	Reverse	Reverse
Controlled variable	$T_{1,12}$	$T_{2,11}$	$T_{3,3}$	$T_{stream14}$
Manipulated variable	Q_{EDC}/F	Q_{RC}/F	RR2	Q_{cooler}
Transmitter range / ℃	0~223.92	0~405.33	0~274.88	0~89.87
Controller output range / (GJ/h)	0~0.096	0~0.055	0~1.11	-42.68~0
K_U	1.6	1.95	6.91	0.26
P_U / min	6.48	7.2	6.0	2.4
K_C	0.50	0.61	2.16	0.08
τ_I / min	14.52	15.84	13.20	5.28

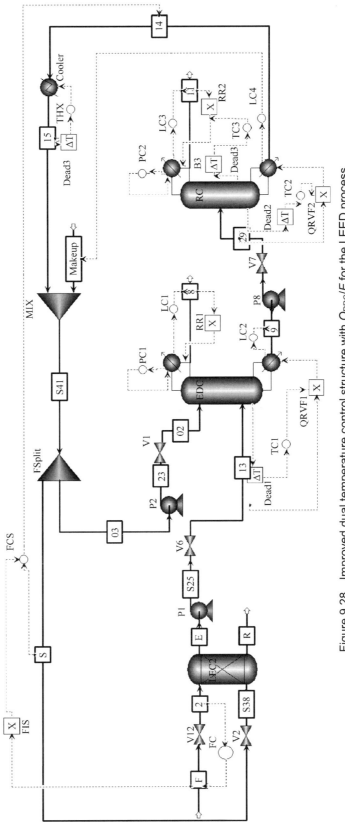

Figure 9.28 Improved dual temperature control structure with Q_{RRC}/F for the LEED process

Chapter 9 Hybrid Process Including Extraction and Distillation

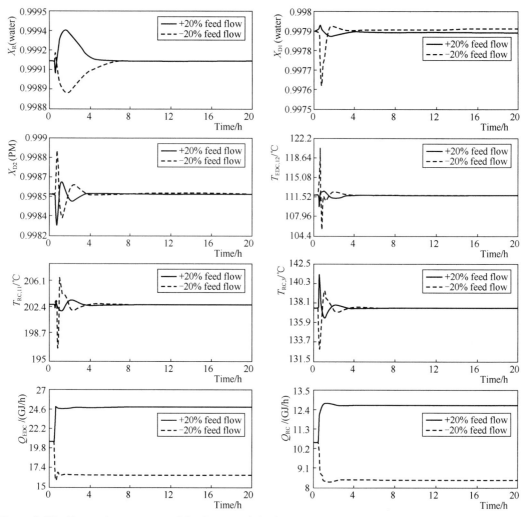

Figure 9.29 Dynamic responses of the improved dual temperature control structure with Q_{RRC}/F for the LEED process after introducing ±20% feed flow rate disturbances

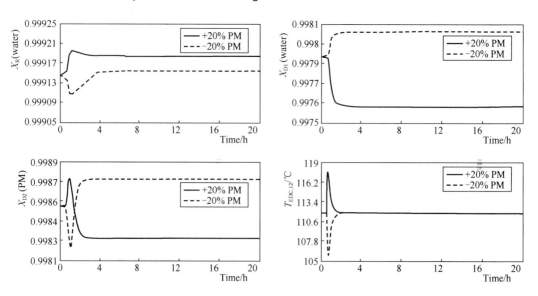

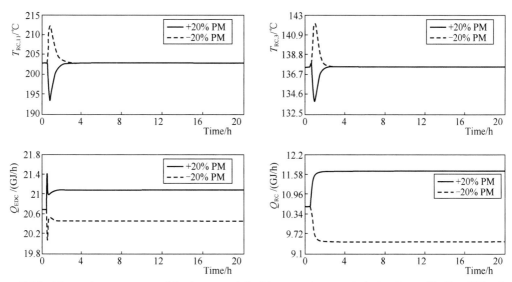

Figure 9.30 Dynamic responses of the improved dual temperature control structure with Q_{RRC}/F for the LEED process after introducing ±20% feed composition disturbances

9.5 Energy-saving Hybrid Process with Mixed Solvent

Figure 9.31 shows the binary interaction parameters between substances. Figure 9.32 shows the optimized process flowsheet for hybrid extraction distillation process with mixed solvent.

Component i	PM	WATER	PM	WATER	PM	CHCL3
Component j	WATER	CHCL3	CHCL3	C8H16-01	C8H16-01	C8H16-01
Temperature units	C	C	C	C	C	C
Source	USER	APV84 LLE-ASPEN	R-PCES	APV84 LLE-ASPEN	R-PCES	R-PCES
Property units						
AIJ	-1.53	8.8436	0	0	0	0
AJI	7.626	-7.3519	0	0	0	0
BIJ	431.045	-1140.12	-474.553	2284.69	-71.0739	511.697
BJI	-1998.75	3240.69	891.344	312.536	79.4968	-262.477
CIJ	0.3	0.2	0.3	0.2	0.3	0.3
DIJ	0	0	0	0	0	0
EIJ	0	0	0	0	0	0
EJI	0	0	0	0	0	0
FIJ	0	0	0	0	0	0
FJI	0	0	0	0	0	0
TLOWER	0	-1	25	25	25	25
TUPPER	1000	54	25	25	25	25

Figure 9.31 Binary interaction parameters between substances

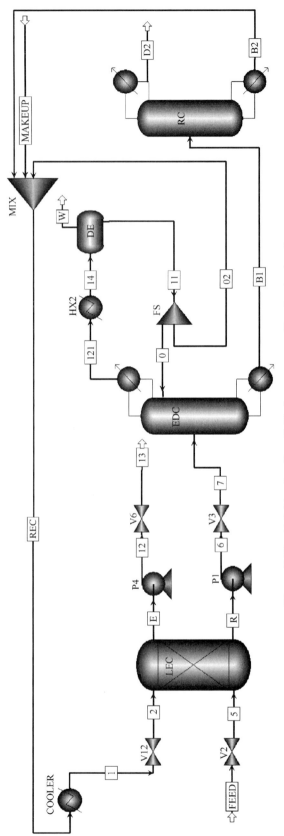

Figure 9.32 Hybrid extraction distillation process with mixed solvent

(1) Inputting the information of feed stream and the extraction stage of liquid-liquid extraction column, as shown in Figure 9.33.

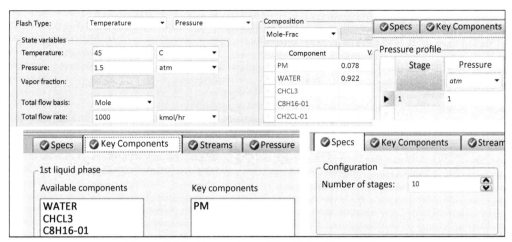

Figure 9.33 Parameters of liquid-liquid extraction

(2) Inputting the parameters of C2 column, as shown in Figure 9.34.

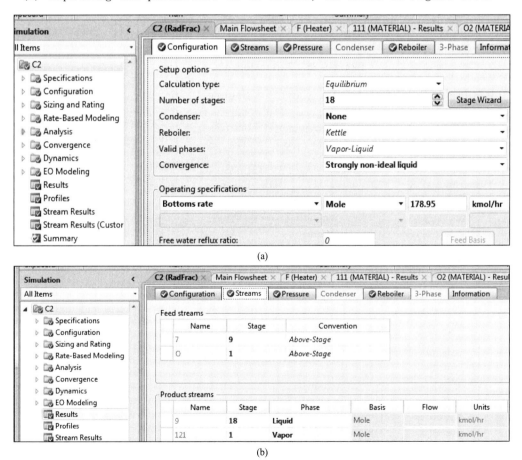

Figure 9.34 Parameters of C2 column

(3) Inputting the parameters of C3 column, as shown in Figure 9.35.

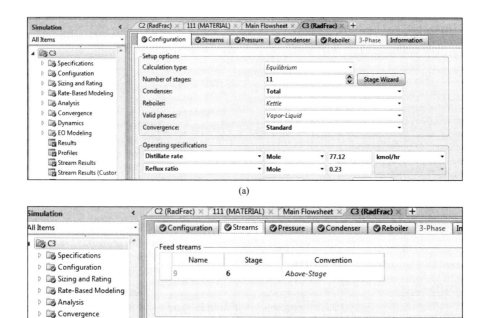

Figure 9.35 Parameters of C3 column

(4) Click C2 | Results, as shown in Figure 9.36.

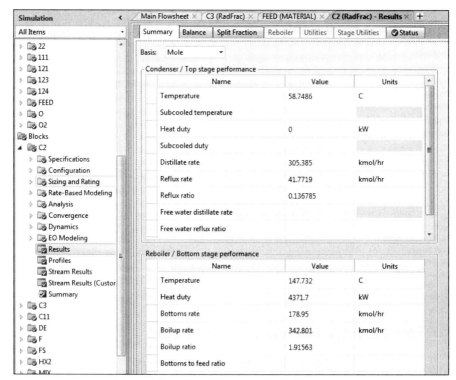

Figure 9.36 Simulation results of C2 column

(5) Click C3 | Results, as shown in Figure 9.37.

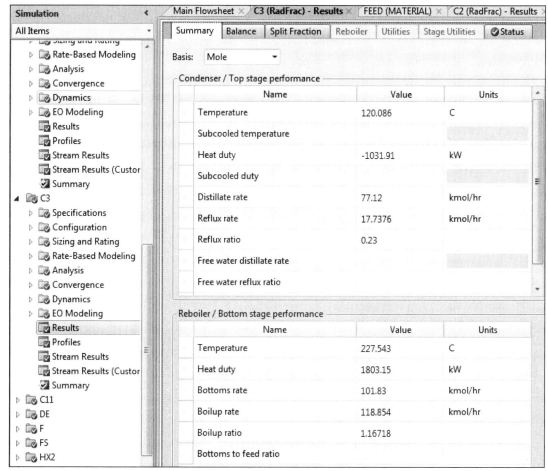

Figure 9.37　Simulation results of C3 column

9.6　Dynamics of Hybrid Process with Mixed Solvent

9.6.1　Selection of Temperature-sensitive Trays

The detail information of binary interaction parameters is shown in Figure 9.31. Before researching the dynamic control characteristics, it is necessary to set some parameters in the steady-state process, such as: reflux tank size, column size, etc. The pressure drop between the distillation trays is set to 0.0068 atm, and the design of the main equipment dimensions is determined by Luyben. The size of reflux tank is set according to the liquid holding capacity which can guarantee 10 minutes. The liquid holding capacity of the reflux tank is calculated according to the flow rate of the gas phase at the top of the column and the flow rate of the liquid phase at the last tray in the column, respectively. In order to achieve efficient separation of the aqueous and organic phases, the size of the phase separator is designed to be larger, based on half the volume of the phase splitter at a liquid hold of 40 min. Since the liquid-liquid extraction column does not support pressure driven, the flow driven is selected to convert the steady-state simulation into a dynamic file. In the flow driven dynamic simulation, the outlet pressure and flow rate are determined by the import conditions and equipment regulations, which are not affected by the pressure of the downstream modules. Therefore, the control of some variables such as liquid level and flow rate is different from pressure driven.

For dynamic control, the choice of temperature-sensitive trays is critical. The slope criterion is the most common method used to select the temperature-sensitive trays of the hybrid process including liquid-liquid extraction and heterogeneous azeotropic distillation. The slope criterion results are shown in Figure 9.38. For column EDC, the temperature slope at the 11th tray is the largest, so it is chosen as the temperature-sensitive tray. For column RC, the 8th tray is chosen as RC temperature-sensitive tray.

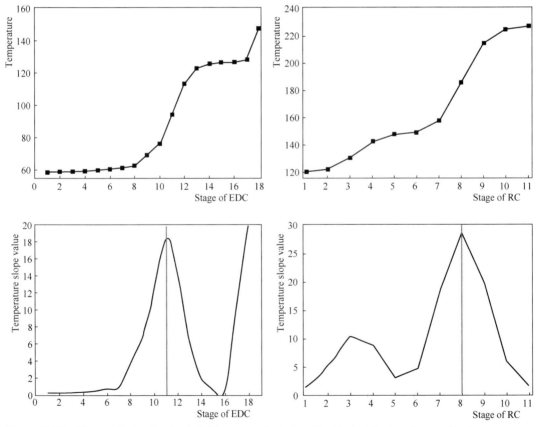

Figure 9.38 Slope criterion for the hybrid process including liquid-liquid extraction and heterogeneous azeotropic distillation

9.6.2 Control Structure with Fixed Reflux Ratio

Section 9.5.1 completes the necessary parameter setting and the selection of temperature-sensitive tray. Then the optimal steady-state simulation of hybrid process with mixed solvent is introduced into Aspen Plus Dynamics, as shown in Figure 9.39.

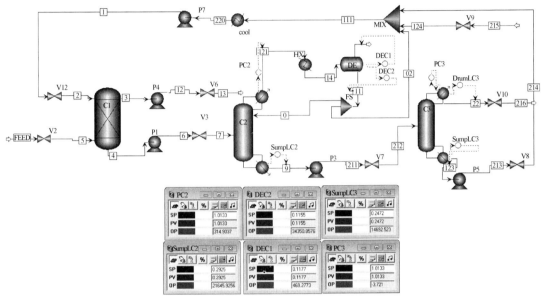

Figure 9.39 Dynamic control for liquid-liquid extraction heterogeneous azeotropic distillation

Everything is ready for relay-feedback test. Click the **Tune** button at the top and right of TC1 controller panel to open the window, as shown in Figure 9.40.

Set the test method to Closed loop ATV, the default value of the relay output amplitude is 5%. For highly non-linear distillation column, the amplitude must be reduced.

Click the **Plot** button above the controller panel to bring up the drawing window, check the dynamic response.

To start testing, first change the mode of operation from "Initialization" to "Dynamic", click the **Run** button, and then click the **Start test** button in the Tune window. After several (5~6) cycles, pause the run and click the **Finish test** button to end the test. The limit gain is 0.57% and the

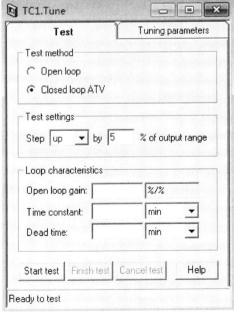

Figure 9.40 TC1 controller Tune window

limit period is 9.71min. Figure 9.41 shows the test results.

The time scale in the figure is quite small. In order to get a good-looking chart, you must reduce the default value of the drawing interval by 0.01 h.

Aspen Dynamics refers to this parameter as communication time. In the toolbar on the top of the Aspen

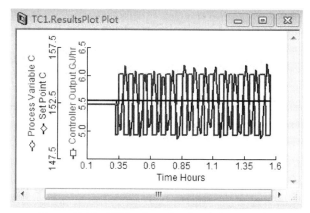

Figure 9.41 TC1 relay - feedback test results

Dynamics window, select run and run options, and communication time can be set in the open window to 0.001 h, as shown in Figure 9.42. In addition to reducing the running speed to a certain extent, this parameter does not affect the results of dynamic simulation, it only changes the shape of the chart.

Finally, click the Tuning parameters page tab and use the default Tyreus-Luyben as the value of the Tuning rule. Click the **calculate** button and the resulting controller is set to gain K_C=0.57 and integration time τ_I=9.72 min. Click the **update controller** button to add them to the controller. Add the infrastructure control structure when there is no problem with the initialization operation. Figure 9.43 shows the basic control structure and control panel of liquid-liquid extraction heterogeneous azeotropic distillation process.

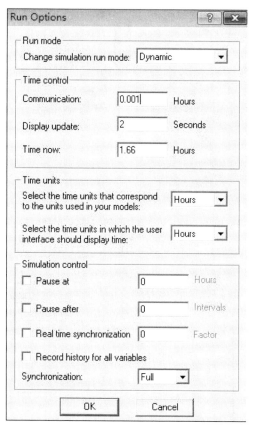

Figure 9.42 Run Options window

Exercises

1. The process of the separation of PM and water process via hybrid extraction-distillation was designed. Chloroform and 2-ethylhexanoic acid were used in the separation of PM and water azeotropic system. Detailed information about the basic control structure is shown in Figure 9.43. Complete the dynamic control of the reboiler heat duty to the feed flow rate.

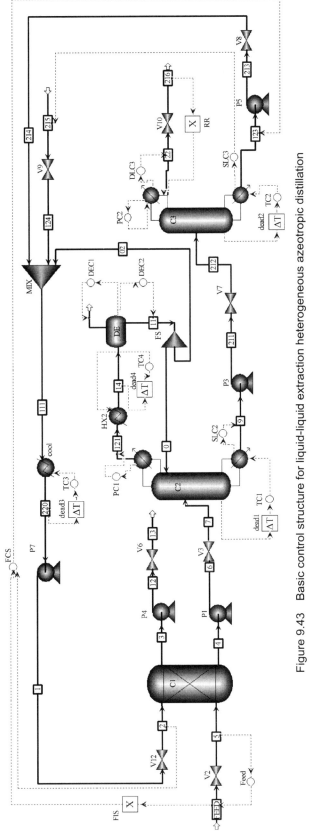

Figure 9.43 Basic control structure for liquid-liquid extraction heterogeneous azeotropic distillation

Chapter 9 Hybrid Process Including Extraction and Distillation

2. In the dynamic of hybrid processes, both the temperature slope and open-loop sensitivity test for the stage of RC have shown that the 3rd stage and 8th stage can be good temperature control trays. Please further explore the dual temperature control structure for the control of RC to search a better control strategy.

References

[1] Chen C, Yu B, Hsu C, et al. Comparison of heteroazeotropic and extractive distillation for the dehydration of propylene glycol methyl ether[J]. Chemical Engineering Research and Design, 2016, 111: 184-195.

[2] Zhao T, Geng X, Qi P, et al. Optimization of liquid-liquid extraction combined with either heterogeneous azeotropic distillation or extractive distillation processes to reduce energy consumption and carbon dioxide emissions[J]. Chemical Engineering Research and Design, 2018, 132: 399-408.

[3] Ma K, Pan X, Zhao T, et al. Dynamic control of hybrid processes with liquid-liquid extraction for propylene glycol methyl ether dehydration[J]. Industrial & Engineering Chemistry Research, 2018, 57: 13811-13820.

[4] Dai Y, Xu Y, Wang S, et al. Dynamics of hybrid processes with mixed solvent for recovering propylene glycol methyl ether from wastewater with different control structures[J]. Separation and Purification Technology, 2018, 229: 115815.

Chapter 10

Batch Distillation Integrated with Quasi-continuous Process

10.1 Introduction

Batch distillation (BD) is widely used in fine chemical and biopharmaceutical industries because of the flexible operation. It can separate the components of the mixture simultaneously. For the separation of azeotropes, pressure-swing batch distillation (PSBD) is designed. The separation of tetrahydrofuran/methanol/water ternary mixture with two azeotropes is done. The feasibility is analyzed based on the residue curve maps. Two pressure-swing batch distillation processes, which are double column batch stripper (DCBS) process and its integration with quasi-continuous process named triple column process are developed. Control schemes are explored to realize stable separation. Based on the minimum Total Annual Cost (TAC), the processes are compared and the advantages for the triple column process are analyzed.

10.2 Feasibility of Pressure-swing Batch Distillation Based on the Ternary Residue Curve Maps

The ternary residue curve maps (RCMs) of tetrahydrofuran/methanol/water at 0.6 atm and 5 atm are shown in Figure 10.1. There are two minimum azeotropes named double azeotropic system, which are tetrahydrofuran/water and tetrahydrofuran/methanol.

At 0.6 atm, the original mixture with composition of 70% tetrahydrofuran, 15% methanol and 15% water locates at Region 1 and the residue curves point to water, so pure water will be obtained at the bottom of the distillation column

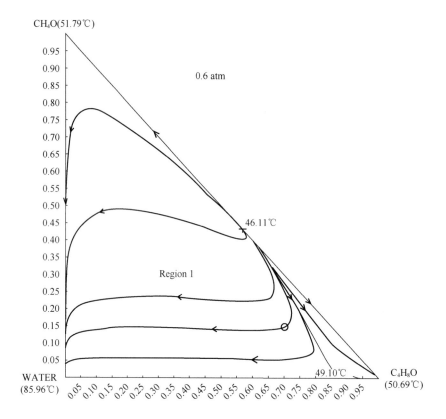
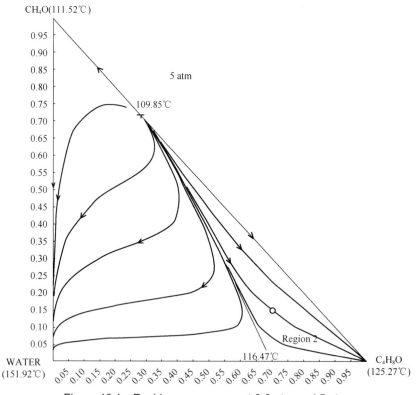

Figure 10.1　Residue curve maps at 0.6 atm and 5 atm

and the mixture will be distilled at the top. At 5 atm, the feed composition locates at Region 2 and the residue curves point to tetrahydrofuran, so pure tetrahydrofuran will be obtained. As for methanol, it can be obtained at low pressure when most water has been separated. At that moment, the mixture in the feed pot is mainly methanol/tetrahydrofuran. The composition of methanol becomes higher with water and tetrahydrofuran removing from the system. At the end of the process, all the components in the ternary system will be separated to high purity.

10.3 Double Column Batch Stripper Process

10.3.1 Design of Double Column Batch Stripper Process

According to the analysis of distillation regions and residue curves in Figure 10.1, a DCBS process is designed to separate the mixture. As shown in Figure 10.2, 100 kmol ternary mixture with the composition of 70% tetrahydrofuran, 15% methanol and 15% water is filled in the feed pot with the volume of 8.66 m^3. Two columns are set to obtain pure components at different operating pressures based on the Radfrac model. The feed streams from the feed pot to the top of the two columns (LPC and HPC) are sent simultaneously.

In the LPC, the high-purity water with composition of 99.9% is obtained and then sent to the water product vessel (V-LP1) whose volume is 11.1 m^3. The methanol can also be obtained from the LPC. The operating pressure of LPC is 0.6 atm and the column diameter is 0.61 m. The stage number of the column is set to 30, as shown in Figure 10.3. In the HPC, the high-purity tetrahydrofuran with composition of 99.9% is withdrawn. The product vessel volume (V-HP) is 24.3 m^3. The operating pressure of HPC is 5 atm and the column diameter is 0.48 m. The stage number is 30, as shown in Figure 10.4. All the vessels are simulated by the Flash 2 model, as shown in Figure 10.5. The volume of the transition vessel is 9 m^3. Methanol is collected in V-LP2.

In the steady-state simulation, the streams F-LPC and F-HPC are disconnected from the V-F, as shown in Figure 10.6. The vessels V-T and V-LP2 are disconnected from LPC, as shown in Figure 10.7.

In the steady-state simulation, mixtures cannot be separated because batch distillation is a dynamic process. When exporting to dynamic process from steady-state, connecting these streams, it can separate the mixtures after running for a few hours.

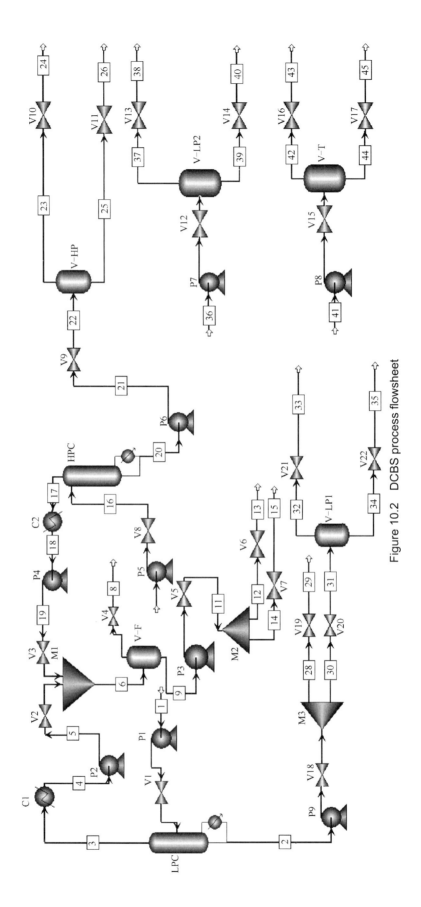

Figure 10.2 DCBS process flowsheet

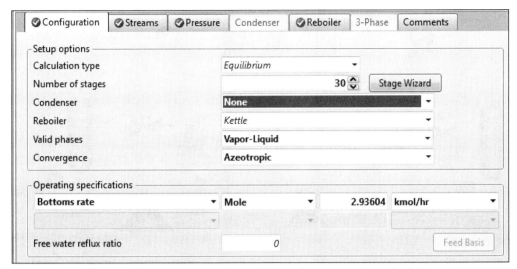

Figure 10.3 The configuration of LPC in the steady-state simulation

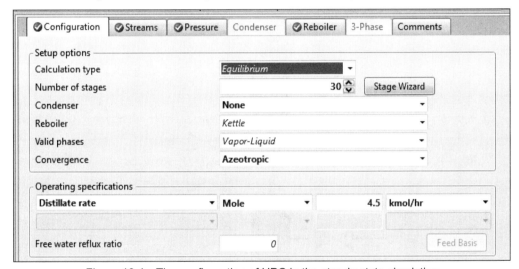

Figure 10.4 The configuration of HPC in the steady-state simulation

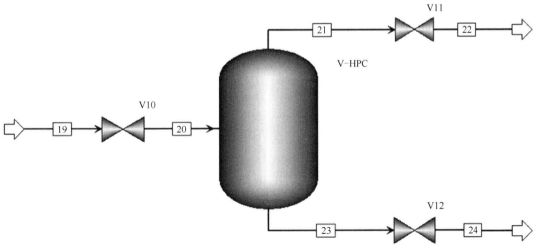

Figure 10.5 Flash 2 model

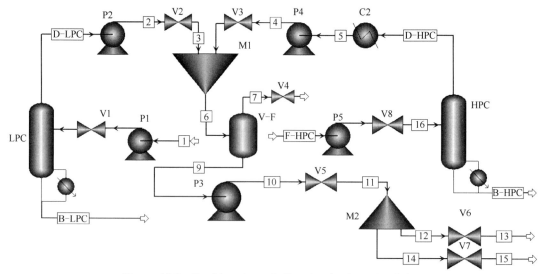

Figure 10.6　Double column in the steady-state simulation

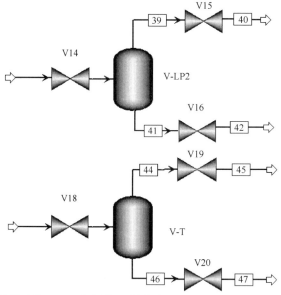

Figure 10.7　The vessels V-T and V-LP2 in the steady-state simulation

10.3.2　Control of Double Column Batch Stripper Process

Pressure controllers are set to keep the operating pressure of the columns stable by manipulating the heat duty of the condensers, as shown in Figure 10.8 and Figure 10.9.

A composition-flowrate cascade control structure is set to keep the tetrahydrofuran composition at 99.9% by manipulating the feed flow rate of HPC. As shown in Figure 10.10, a dead time of 3 min is added in the composition control loop. A flow controller is set to control the feed flow rate of LPC by manipulating the opening of the valve, as shown in Figure 10.11. The parameters of the FC1 are shown in Figure 10.12.

双塔间歇精馏工艺在动态模拟中的操作步骤介绍

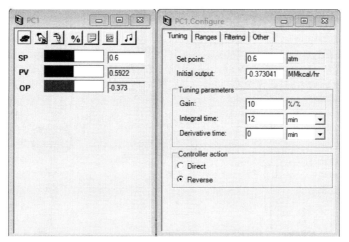

Figure 10.8　The parameters of the PC1

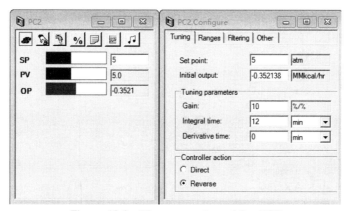

Figure 10.9　The parameters of the PC2

Figure 10.10　The parameters of the FC2 and CC1

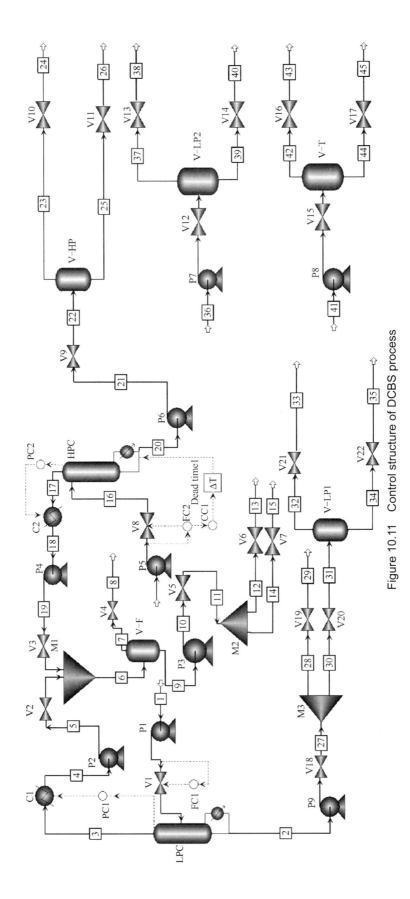

Figure 10.11 Control structure of DCBS process

198　Chemical Process Simulation

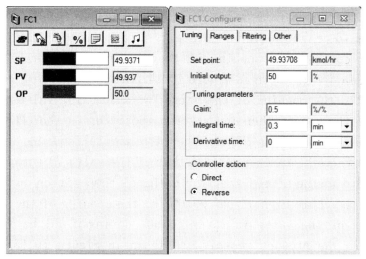

Figure 10.12　The parameters of the FC1

　　The end of the process is at the moment when the mixture in the feed pot is separated. Before that, the feed streams of the columns are fed simultaneously, so the hydraulic states on the column stages could keep normal. Figure 10.13 shows the changes of compositions in the bottom stream of LPC. The tetrahydrofuran content keeps at a low value which means that the component is all distilled at the top. At the first 10 h, the water content is very high and it is collected in the V-LP1. At the following 4 h, the water content drops dramatically while the methanol content increases. During this period, the purity is low for both water and methanol. Hence, a transition vessel should be set to ensure that the purity of the products is high enough in their product vessels, because LPC is used to separate water and methanol in turn.

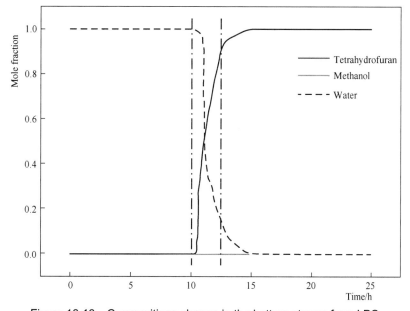

Figure 10.13　Compositions change in the bottom stream from LPC

LPC is used to separate water first and then methanol. Therefore, the sequence needs to be controlled accurately. A task was compiled in the dynamic simulation to control the opening of the valves at different moments. The detailed information is shown in Figure 10.14. In the process of producing water (before 9.12 h), the opening of the valve V1 is 50%. The valves V2 and V3 are off. 9.12 h is the moment when the water composition in the LPC bottom stream begins to become lower than 99.9%. When the simulation runs at 9.12 h, the opening of the valve V1 turns to 0% and that of the valve V2 turns to 50% during 0.001 h. 13.44 h is the moment when the methanol composition in the LPC bottom stream begins to become higher than 99.0%, so the opening of the valve V2 turns to 0% and that of the valve V3 turns to 50% during 0.001 h. The changes of compositions in the product vessels are shown in Figure 10.15. It should be noticed that the products accumulate in the vessels during the whole process, so the final compositions are marked by yellow circles in the figure. The purity of water and tetrahydrofuran are up to 99.9% and 99.85% while the purity of methanol is only 99.4%.

```
Task - a *
1   — Task a // <Trigger>
2   // For event driven tasks, <Trigger> can be one of:
3   //    Runs At <time>          e.g. Runs At 2.5 or
4   //    Runs When <condition>   e.g. Runs When bl.y >= 0.6 or
5   //    Runs Once When <condition> e.g. Runs Once When bl.y >= 0.6
6   // Ramp  (<variable>, <final value>, <duration>, <type>);
7   // SRamp (<variable>, <final value>, <duration>, <type>);
8   // Wait For <condition> e.g. when bl.y < 0.6;
9   // (Use Wait For to stop the task firing again once trigger condition has been met)
10  runs at 9.12
11  ramp (blocks ("V1").Pos,0,0.001);
12  ramp (blocks ("V2").Pos,50,0.001);
13  End
14
```

(a)

```
Task - b *
1   — Task b // <Trigger>
2   // For event driven tasks, <Trigger> can be one of:
3   //    Runs At <time>          e.g. Runs At 2.5 or
4   //    Runs When <condition>   e.g. Runs When bl.y >= 0.6 or
5   //    Runs Once When <condition> e.g. Runs Once When bl.y >= 0.6
6   // Ramp  (<variable>, <final value>, <duration>, <type>);
7   // SRamp (<variable>, <final value>, <duration>, <type>);
8   // Wait For <condition> e.g. when bl.y < 0.6;
9   // (Use Wait For to stop the task firing again once trigger condition has been met)
10  runs at 13.44
11  ramp (blocks ("V2").Pos,0,0.001);
12  ramp (blocks ("V3").Pos,50,0.001);
13  End
14
```

(b)

Figure 10.14 Detailed tasks at (a) 9.12 h and (b) 13.44 h

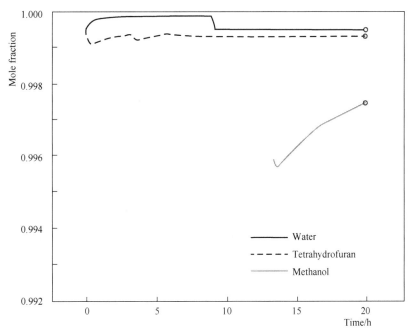

Figure 10.15 Changes of compositions in product vessels

10.4 Triple Column Process

10.4.1 Design of Triple Column Process

In the DCBS process, the methanol purity is 99.4% which is much lower than the purity demand of superior product. Therefore, a triple column process was designed to further improve the separation effect. A third column was set in the DCBS process to separate the methanol/water mixture directly. As shown in Figure 10.16, when the methanol/water mixture from the bottom stream of LPC is sent to the third column, the feed stage of the column is not set at the bottom or the top and both pure components of the binary mixture can be obtained simultaneously instead of sequentially. However, the feed stream of the third column is from a pressure-swing batch distillation. Hence, the operation of the third column has the characteristics of continuous distillation and batch distillation. The pressure-swing triple column process can be also called quasi-continuous process.

Figure 10.17 shows the configuration of the triple column process. The bottom stream from LPC is divided into two substreams. One of them is connected with the V-LP1 while the other is sent to the third column.

In the steady-state simulation, LPC and HPC are the same as in DCBS process, but TC is disconnected from LPC.

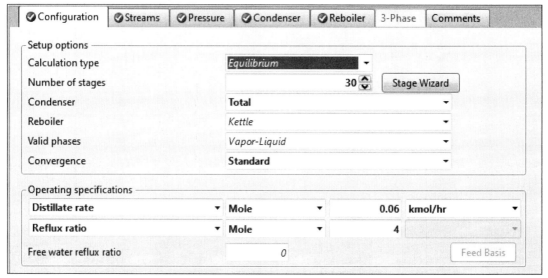

Figure 10.16 The configuration of TC in the steady-state simulation

10.4.2 Control of Triple Column Process

At the first 9.12 h, the bottom stream is only sent to the V-LP1 because there is high-purity water in the stream. After that, the stream turns to the third column to implement these parathion of methanol and water until the process finishes. Water is withdrawn at the bottom and then sent to V-LP1 while methanol is distilled at the top and sent to V-LP2.

Based on the control structure of DCBS process, the controllers for the third column are shown in Figure 10.18.

The operating pressure is controlled by manipulating the heat duty of the condenser, as shown in Figure 10.19. The parameters of the CC2 and CC3 are shown in Figure 10.20. The flow rate of the streams, which flow out from the top and the bottom of the column, is controlled. The task is also compiled in the simulation.

Figure 10.21 shows the changes of compositions in the product vessels. The tetrahydrofuran mole fraction fluctuates at the first 6 h and then becomes stable. For water, the opening of the valves is changed at 9.12 h. Therefore, the water purity increases initially and then drops because of the addition of methanol. After 9.12 h, it becomes stable. While for methanol, the mole fraction is unchanged because there is no stream entering to the vessel before 9.12 h. Then the composition increases with the process proceeding. The final compositions of the products in the three vessels are 99.9% for methanol, 99.9% for water and 99.85% for tetrahydrofuran.

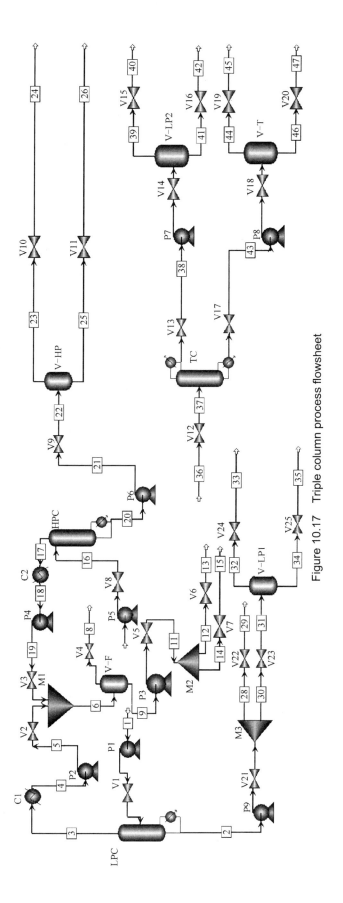

Figure 10.17　Triple column process flowsheet

Chapter 10　Batch Distillation Integrated with Quasi-continuous Process

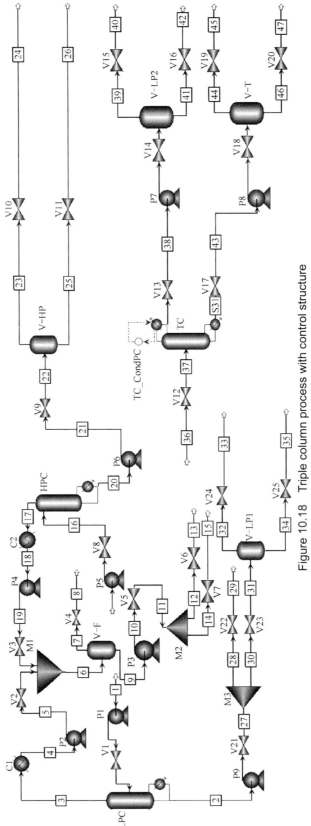

Figure 10.18 Triple column process with control structure

204 Chemical Process Simulation

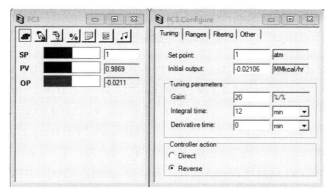

Figure 10.19 The parameters of the PC3

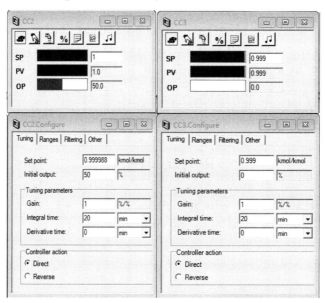

Figure 10.20 The parameters of the CC2 and CC3

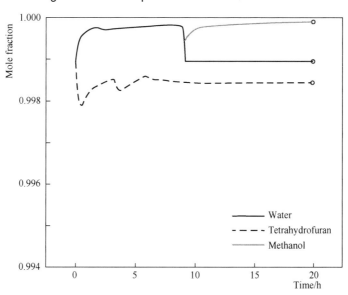

Figure 10.21 Changes of compositions in products vessels

Exercises

1. Pressure-swing batch distillation was investigated to separate the binary azeotrope of dichloromethane/methanol. In the steady-state simulation with UNIQUAC property method, 100.0 kmol binary mixture with the composition of 83.0% dichloromethane and 17.0% methanol was fed into the feed pot. The purity of the separated dichloromethane and methanol is higher than 99.8%. Complete the steady-state simulation and dynamic simulation.

2. The middle vessel batch distillation process for separating the methyl formate/methanol/water ternary system was investigated using Aspen Plus and Aspen Plus Dynamics. The flow rate of feed stream that was fed to the sump of the LS column was set at 1000.0 kmol/h, and its composition was 30.0% (mole fraction) methyl formate, 30.0% methanol, and 40.0% water. The purity of all components is higher than 99.8%. Complete the steady-state simulation and dynamic simulation.

References

[1] Klein A, Repke J U. Regular and inverted batch process structures for pressure swing distillation: a case study[J]. Asia - Pacific Journal of Chemical Engineering, 2009, 4(6): 893-904.

[2] Li X, Yang X, Wang S, et al. Separation of ternary mixture with double azeotropic system by a pressure-swing batch distillation integrated with quasi-continuous process[J]. Process Safety and Environmental Protection, 2019, https://doi.org/10.1016/j.psep.2019.05.040.

[3] Luyben W L. Aspen Dynamics simulation of a middle-vessel batch distillation process[J]. Journal of Process Control, 2015, 33: 49-59.

[4] Li X, Zhao Y, Qin B, et al. Optimization of pressure-swing batch distillation with and without heat integration for separating dichloromethane/methanol azeotrope based on minimum total annual cost[J]. Industrial & Engineering Chemistry Research, 2017, 56(14): 4104-4112.

[5] Zhu Z, Li X, Cao Y, et al. Design and control of a middle vessel batch distillation process for separating the methyl formate/methanol/water ternary system[J]. Industrial & Engineering Chemistry Research, 2016, 55(10): 2760-2768.

Chapter 11

Simulation of Chemical Reaction Process Based on Reaction Kinetics

11.1 Introduction

Future development of the chemical industry must be based on green chemistry and green engineering to avoid the social and environmental problems currently associated with this industry. Alleviating the environmental pollution and high energy consumption caused by the chemical industry are major challenges for its development. However, most industrial chemical production processes can be made cleaner by optimizing the process operating conditions based on thermodynamic and economic performance analyses.

Chemical reaction process is the core of the chemical conversion technology so that improving chemical reaction process efficiency is an important task for scientists and engineers. This chapter mainly introduces the application of Aspen Plus in the simulation, optimization and dynamic control of chemical reactor.

Aspen Plus provides seven different reactor modules according to different reactor types, as shown in Table 11.1.

Table 11.1 Introduction of reactor unit module

Module	Illustrate	Funcation
Rstoic	Stoichiometric reactor	The reactor module that simulates a known degree of reaction or conversion.
Ryield	Yield reactor	The reactor module that simulates a known yield.
Requil	Equilibrium reactor	The chemical equilibrium and phase equilibrium are calculated by the stoichiometric relationship of chemical reaction formula.
RGibbs	Gibbs reactor	The chemical equilibrium and phase equilibrium are calculated by the minimizing Gibbs free energy.
RCSTR	Continuously stirred tank reactor	Simulated continuously stirred tank reactor.
RPlug	Plug flow reactor	Simulated plug flow reactor.
RBatch	Batch reactor	Simulated batch or semi-batch reactor.

Reactor module can be divided into three categories:

(1) The reactor based on material balance includes stoichiometric reactor (Rstoic) module and yield reactor (Ryield) module.

(2) The reactor based on chemical equilibrium includes equilibrium reactor (Requil) module and Gibbs (RGibbs) reactor module.

(3) The dynamic reactor includes continuously stirred tank reactor (RCSTR) module, plug flow reactor (RPlug) module and batch reactor (RBatch) module.

For any reactor module, there is no need to input the reaction heat. Aspen Plus calculates the reaction heat according to the heat of formation. For the kinetic reactor module, the stoichiometric relationship of the reaction and the data of the reactor module are defined by the chemical reaction function.

11.2 Continuously Stirred Tank Reactor

The reactor module of continuously stirred tank reactor strictly simulates continuous stirred tank reactor, which can simulate single-phase, two-phase or three-phase system. RCSTR module assumes that the reactor is completely mixed, that is, the reactor has the same properties and composition as the outlet streams, which can handle both the kinetic and equilibrium reactions and the reactions with solids. Users can provide reaction kinetics through built-in reaction models or user-defined subroutines.

RCSTR module needs to specify reactor pressure, temperature or heat duty, effective phase state, reactor volume or residence time, etc. If the RCSTR module links two or three outlets, the outlet phase of each stream should be set in the Stream page. The chemical reaction formula in RCSTR module is defined by the chemical reaction.

If the volume of the reactor is determined, the residence time of the RCSTR is calculated by the following formulas:

total residence time: $RT = \dfrac{V_R}{F\sum f_i V_i}$ (11-1)

i phase residence time: $RT_i = \dfrac{V_{pi}}{F f_i V_i}$ (11-2)

Where, RT is total residence time; RT_i is i phase residence time; V_R is reactor volume; F is the total molar flow rate (outlet); V_{pi} is i phase volume; V_i is i phase molar volume; f_i is i phase mole fraction.

Next, the application of RCSTR module is introduced through cyclohexanone ammoximation process.

11.3 Simulation of Cyclohexanone Ammoximation Process

The cyclohexanone ammoximation process plays an important role in the production of caprolactam. Unlike other industrial processes, the liquid phase ammoximation of cyclohexanone oxime can be completed in one step without involving hazardous chemicals or by-products.

In this section, the effects of the reaction temperature, space time and raw material ratio on the cyclohexanone oxime yield are explored. Dynamic control of the cyclohexanone ammoximation process is explored under the optimized conditions. The influences of feed composition and feed flow rate disturbances on industrial production are analyzed.

11.3.1 Steady-state Simulation of Cyclohexanone Ammoximation Process

The cyclohexanone ammoximation process uses tert-butyl alcohol ($C_4H_{10}O$) as solvent. The basic flowsheet design for $C_6H_{11}NO$ production is shown in Figure 11.1. A mixture of ammonia (NH_3), hydrogen peroxide (H_2O_2) and cyclohexanone ($C_6H_{10}O$) enters the RCSTR and reacts under the action of a TS-1 catalyst to form $C_6H_{11}NO$ directly. The reactor outlet stream is a multicomponent mixture

简单反应动力学反应器的模拟

that requires further separation. The crude $C_6H_{11}NO$ solution obtained from the synthesis unit enters decanter D1 after the removal of $C_4H_{10}O$, oxygen and NH_3 by separators S1 and S2. The C_7H_8–$C_6H_{11}NO$ mixture flows into distillation column C1. The water phase enters extraction column C2 from stream 6. M3 is a storage tank for the solvent C_7H_8. The extracted C_7H_8–$C_6H_{11}NO$ mixture is refluxed into decanter D1, and the raffinate is discharged at the bottom of extraction column C2. C_7H_8 is distilled from the top of column C1 and recovered to the C_7H_8 storage tank. Pure $C_6H_{11}NO$ solvent flows from the bottom of column C1 and enters the product storage tank.

The intrinsic kinetic equation of the cyclohexanone ammoximation and H_2O_2 decomposition reaction is obtained:

$$\bigcirc\!\!=\!O \; + \; NH_3 \; + \; H_2O_2 \; \xrightarrow{TS-1} \; \bigcirc\!\!=\!NOH \; + \; H_2O \quad (11-3)$$

For this reaction process, the kinetic parameters of the reaction must be set in the Aspen reactor module. For the simple kinetics simulation of the cyclohexanone ammoximation reaction, a reaction kinetics simulation of the main reaction and the decomposition of hydrogen peroxide are carried out without considering other side reactions. The POWERLAW function kinetics equation is selected. The kinetics model is as follows:

$$r_A = -\frac{dC_A}{dt} = k_1 C_A^\alpha C_B^\beta C_C^\gamma \quad (11-4)$$

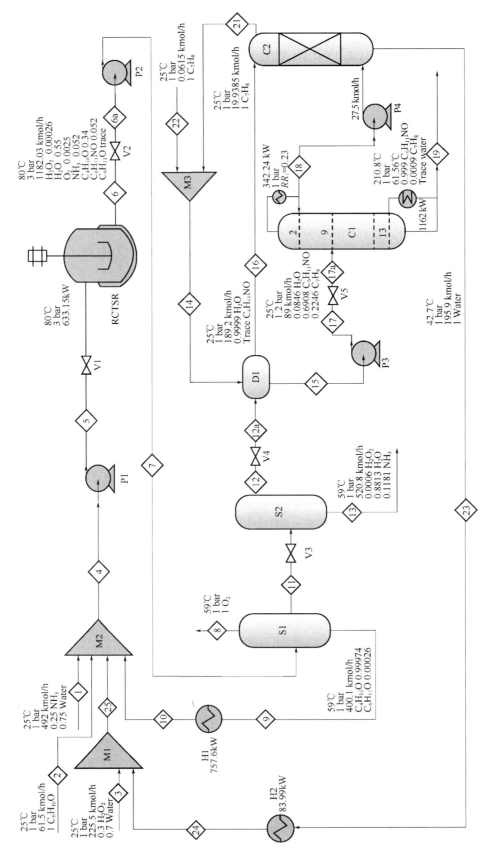

Figure 11.1 Optimal process flow diagram for $C_6H_{11}NO$ production

$$k = k_0 \exp\left(-\frac{E_0}{RT}\right) \tag{11-5}$$

The RCSTR is selected as the cyclohexanone ammoximation reactor, and the kinetic parameters involved in the main and side reactions are taken from the literature. The powerlaw function kinetics equation of the main reaction of cyclohexanone ammine oxidation is as follows:

$$r_A = -\frac{dC_A}{dt} = 4.35 \times 10^{13} \exp\left(-\frac{5.84 \times 10^4}{RT}\right) C_A^{0.90} C_B^{0.12} C_C^{0.81} \tag{11-6}$$

The reaction rate equation for the decomposition of hydrogen peroxide is as follows:

$$r_B = 2.12 \times 10^{13} \exp\left(-\frac{1.12 \times 10^5}{RT}\right) C_B^{1.43} \tag{11-7}$$

Where r_A and r_B represent the reaction rate [mol/(L·min)] of the main reaction and side reaction, respectively; C_A, C_B and C_C represent the concentrations (mol/L) of cyclohexanone, hydrogen peroxide and ammonia, respectively; R represents the gas constant [J/(mol·K)]; T represents the reaction temperature (K). In this way, the preexponential factor and activation energy constant of the reaction kinetics could be set in Aspen Plus. For the basic parameters of the reactor, a reaction temperature of 353.15 K and a space time of 2 hours are selected as the basic reaction conditions.

The simulation steps for this example are as follows:

(1) Start Aspen Plus, enter **File | New | User**, Select **template of General** with **Metric Units**. Saving the file as Example-RCSTR.bkp.

(2) Click **Next**, enter **Methods | Specifications | Global** page. Select **RK-SOAVE** as the property method, as shown in Figure 11.2.

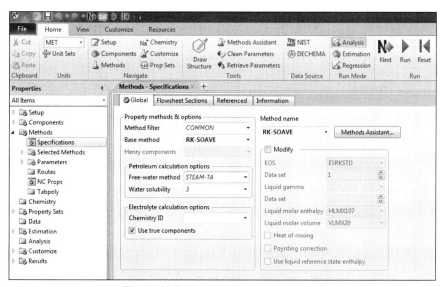

Figure 11.2 Inputting property method

(3) Enter **Properties | Components | Specifications** page and input the relevant components, as shown in Figure 11.3.

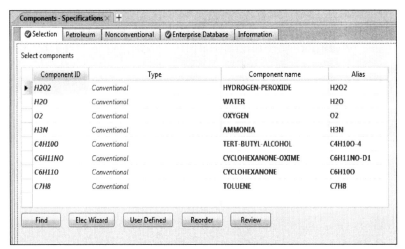

Figure 11.3　Inputting the relevant components

(4) Click **Next**, enter **Blocks | RCSTR | Specifications** page and input the parameters of RCSTR module. The operating pressure is 3 atm. The operating temperature is 353.15 K. Valid phases is Liquid-Only. Residence time is 3 h, as shown in Figure 11.4.

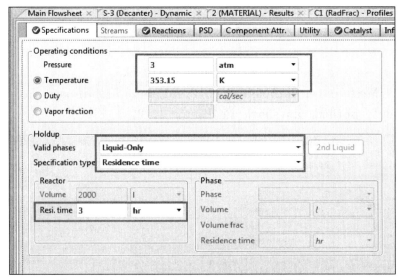

Figure 11.4　Inputting the parameters of RCSTR module

(5) Enter **Reactions | Reactions** page and create chemical reactions. Click **New...** button, Create New ID dialog box appears, the default ID is R-1. Select Type is POWERLAW, as shown in Figure 11.5.

(6) Click **OK**, enter **Reactions | R-1 | Input | Stoichiometry** page. Click **New...** button. Edit Reaction dialog box appears. Reaction type is Kinetic. Input chemical reaction equation 1, as shown in Figure 11.6.

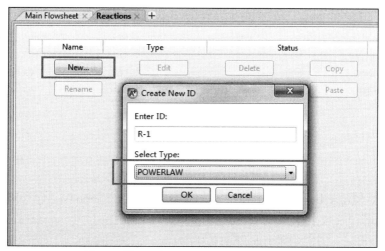

Figure 11.5 Creating chemical reactions

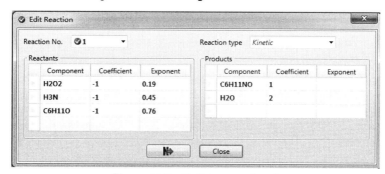

Figure 11.6 Defining reaction 1

(7) Click **Close** button, return **Reactions | R-1 | Input | Stoichiometry** page. Enter **Reactions | R-1 | Input | Kinetic** page, the default is reaction 1. Reacting phase is Liquid. $k=4.351\times10^{13}$, $E=13948.6$ cal/mol, [Ci] basis is Molarity, as shown in Figure 11.7.

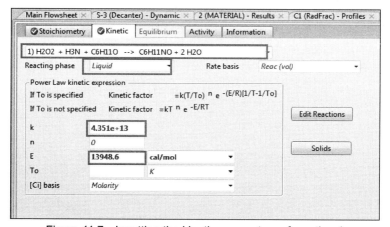

Figure 11.7 Inputting the kinetic parameters of reaction 1

(8) Similarly, inputting the chemical reaction equation 2, as shown in Figure 11.8.

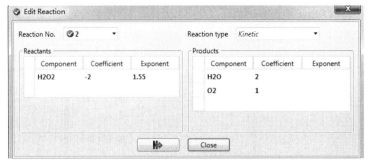

Figure 11.8　Defining reaction 2

(9) Click **Close** button, return **Reactions | R-1 | Input | Stoichiometry** page. Enter **Reactions | R-1 | Input | Kinetic** page, the default is reaction 2. Reacting phase is Liquid. $k=2.12\times10^{13}$, $E=26511.9$ cal/mol, [Ci] basis is Molarity, as shown in Figure 11.9.

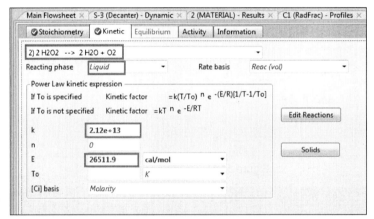

Figure 11.9　Inputting the kinetic parameters of reaction 2

Chemical reaction creation is completed.

(10) Click **Next**, enter **Blocks | RCSTR | Setup | Reactions** page, select **R-1** from Available reaction sets into Selected reaction sets, as shown in Figure 11.10.

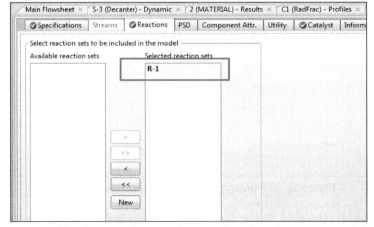

Figure 11.10　Selecting the chemical reaction object in RCSTR module

(11) Click **Next**, Required Input Complete dialog box appears. Click **OK** and run the simulation, as shown in Figure 11.11.

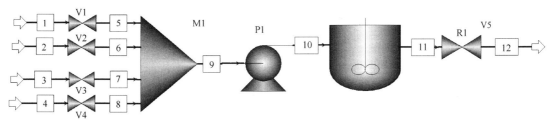

Figure 11.11 Simulation processes of cyclohexanone ammoximation production

(12) Enter **Model Analysis Tools | Sensitivity** page. Click the **New...** button. Using the default ID **S-1**. Create sensitivity analysis module, as shown in Figure 11.12.

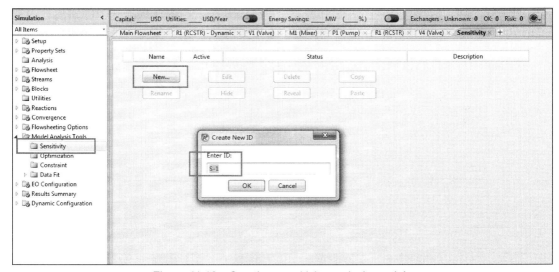

Figure 11.12 Creating sensitivity analysis module

(13) Click **Next**, enter **Model Analysis Tools | Sensitivity | S-1 | Input | Vary** page, and define manipulation variables. What needs to be changed in this case is the reaction temperature of the reactor. Variable range and step size need to be specified. In this case, the range of manipulation variables varies within 325~380 K, the step length is 5 K, as shown in Figure 11.13.

(14) Click **Next**, enter **Model Analysis Tools | Sensitivity | S-1 | Input | Define** page. Define the acquisition variable PY, which refers to the mole fraction of $C_6H_{11}NO$ in product, as shown in Figure 11.14.

(15) Click **Next**, enter **Model Analysis Tools | Sensitivity | S-1 | Input | Tabulate** page. Define the column positions of variables or expressions in the result list, as shown in Figure 11.15.

(16) Click **Next**, pop-up Required Input Complete dialog box. Click **OK** and run simulation. The process is convergent.

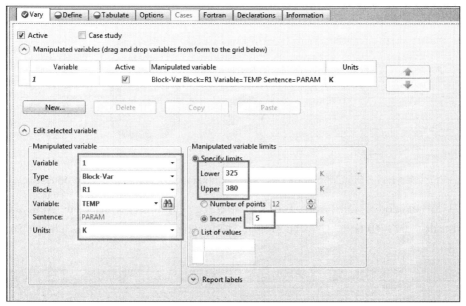

Figure 11.13　Defining manipulation variables

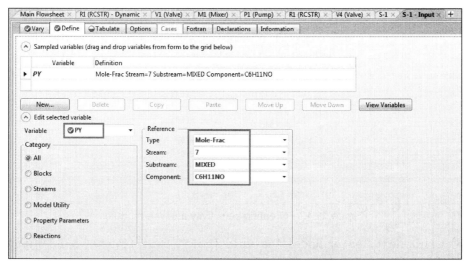

Figure 11.14　Defining the acquisition variable PY

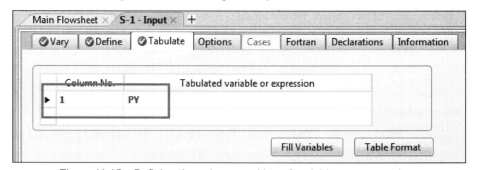

Figure 11.15　Defining the column position of variables or expressions

(17) Enter Model Analysis Tools | Sensitivity | S-1 | Results | Summary page. View the sensitivity analysis results, as shown in Figure 11.16.

Summary | Define Variable | Status

Row/Case	Status	VARY 1 R1 PARAM TEMP K	PY
1	OK	325	0.0498828
2	OK	330	0.0508288
3	OK	335	0.0513611
4	OK	340	0.0516561
5	OK	345	0.0518191
6	OK	350	0.0519095
7	OK	355	0.0519601
8	OK	360	0.0519887
9	OK	365	0.052005
10	OK	370	0.0520144
11	OK	375	0.0520198
12	OK	380	0.052023
13	OK	353.15	0.0519447

Figure 11.16 Sensitivity analysis results

In order to observe the change of product yield and reactor temperature more intuitively, the results are plotted, as shown in Figure 11.17.

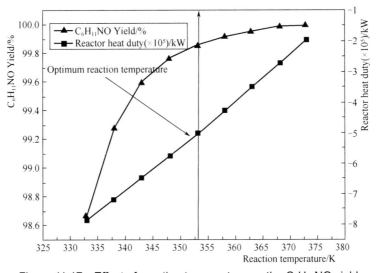

Figure 11.17 Effect of reaction temperature on the $C_6H_{11}NO$ yield

In order to improve the purity of the product, it is necessary to further purify the product. As shown in Figure 11.18, the separation and purification unit of cyclohexanone ammoximation process is added. Input the calculated parameters | Enter the simulation processes of cyclohexanone ammoximation production.

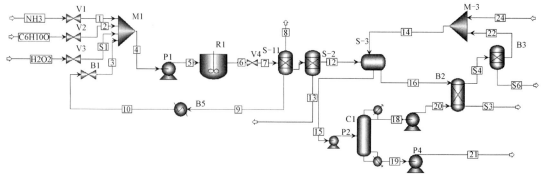

Figure 11.18 Inputting the calculated parameters | Simulation processes of cyclohexanone ammoximation production

The cyclohexanone oxime distillation column is designed. As shown in Figure 11.19, the type of Design specification is Mole purity. The target of specification is 0.999.

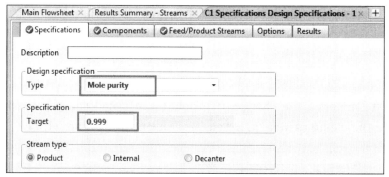

Figure 11.19 Adding design specifications

All stream results are shown in Figure 11.20. It can be seen that the mole fraction of cyclohexanone oxime is 0.999.

	NH3	H2O2	C6H10O	8	13	21	S3	S6	9
C7H8	0	0	0	0	0	0.0615507	9.08846e-06	0	0
Mole Frac									
H2O2	0	0.3	0	0	0.00338375	0	0	0	0
H2O	0.75	0.7	0	0	0.878626	1.7256e-07	1	0.999725	0
O2	0	0	0	1	0	0	0	0	0
H3N	0.25	0	0	0	0.11799	0	0	0	0
C4H10O	0	0	0	0	0	0	0	0	1
C6H11NO	0	0	0	0	0	0.999	1.2655e-11	0.000275538	0
C6H11O	0	0	1	0	0	0	0	0	4.
C7H8	0	0	0	0	0	0.000999828	4.64906e-08	0	0
Mass Flow kg/hr									
H2O2	0	2301.09	0	0	59.992	0	0	0	0
H2O	6647.64	2843.72	0	0	8250.38	0.000191377	3521.81	14.0625	0
O2	0	0	0	70.178	0	0	0	0	0
H3N	2094.76	0	0	0	1047.38	0	0	0	0
C4H10O	0	0	0	0	0	0	0	0	29
C6H11NO	0	0	0	0	0	6959.29	2.79947e-07	0.0243452	0
C6H11O	0	0	6035.91	0	0	0	0	0	1.
C7H8	0	0	0	0	0	5.67132	0.000837415	0	0

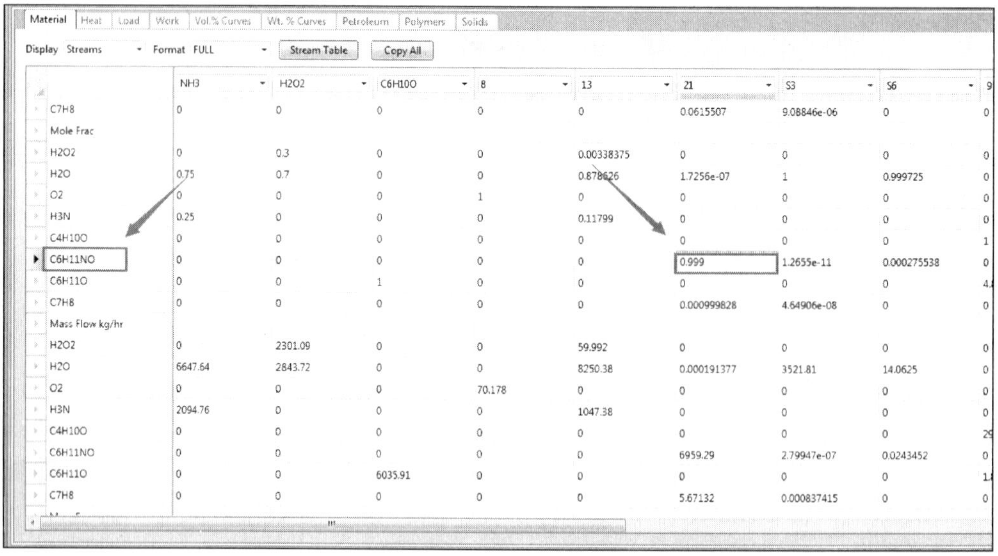

Figure 11.20 Viewing the results of each stream

11.3.2 Dynamic Simulation of Cyclohexanone Ammoximation Process

Aspen Dynamics V8.4 software is used to explore the cyclohexanone ammoximation process. The size of the main equipment is taken from Luyben. The distillation column tray pressure drop is set to 0.0068 atm, and the diameter is 0.91 m. The temperature-sensitive tray for the distillation column is selected using the slope criterion method. Figure 11.21 shows the temperature slope and the profiles of temperature. The 11th stage of the distillation column is selected as the temperature-sensitive stage.

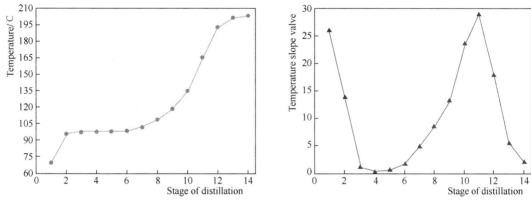

Figure 11.21 Slope of temperature distribution profiles for distillation

The integral time (τ_I) and gain (K_C) of the flow rate controllers are τ_I = 0.3 min and K_C = 0.5, respectively. The level controllers are set to 9999 min and 2, respectively. The two dead-time values for the temperature controllers are 1 minute. Tyreus-Luyben tuning is used to determine the ultimate gain and time. The relevant parameters are shown in Table 11.2.

Table 11.2 Transmitter ranges, controller output ranges, and tuning parameters of three temperature controllers for cyclohexanone ammoximation process

Variable	TC1	TC2	HE
Controlled variable	—	T_{11}	—
Manipulated variable	Q_R	Q_R	Q_H
Transmitter range / K	303.15~403.15	273.15~605.01	273.15~314.97
Controller output range / (GJ/h)	-38.98~38.98	0~8.03	-5.43~0
Gain K_C	5	0.48	0.21
Integral time τ_I / min	6	15.84	5.28

The control structure of the cyclohexanone ammoximation process is shown in Figure 11.22.

Feed flow rate disturbances of ±20% are introduced to the control structure system after 1 h and finished at 20 h. The dynamic responses are shown in Figure 11.23. The heat duty changes in accordance with the feed flow rate disturbances and reaches new stable values within a short time. The purity of the cyclohexanone ammoximation recovers its initial value within 3 hours. Even for large disturbances, the temperature change in the reactor is very small, as is required in industrial production. For the distillation column, the temperature of the sensitive tray also recovers its initial value within 3 hours.

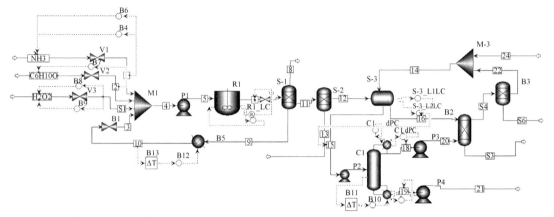

Figure 11.22   The control structure of the cyclohexanone ammoximation process

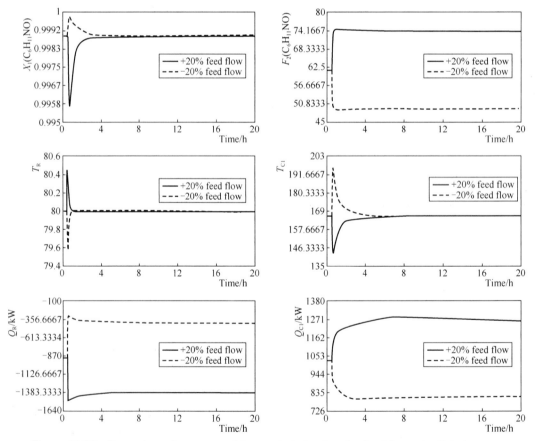

Figure 11.23   Dynamic performances of the control structure for feed flow rate disturbances

In the reaction process, the requirements for the reactant composition are strict. Thus, feed composition disturbances of ±5% are investigated, the dynamic responses are shown in Figure 11.24~Figure 11.27. In addition, ±10% and ±20% feed composition disturbances are investigated. The process could be effectively controlled when the NH_3 composition is varied, but process control could not be achieved for the variations in H_2O_2 and $C_6H_{10}O$. Even in the presence of a large NH_3 fluctuation, the controller could achieve effective control, this finding could guide industrial practice. For −5% feed composition disturbances of H_2O_2 and NH_3, the purity of the product first decreased and then increased over time. For +5% feed composition disturbance of $C_6H_{10}O$, due to the exothermic characteristic of the reaction, the temperature of the reactor will rise instantaneously and then return to its initial value within a short time. For the ±20% NH_3 feed composition disturbances, trends in the variations in temperature, purity, and heat duty are the same as for the 5% feed composition disturbances of NH_3. The dynamic responses are shown in Figure 11.27.

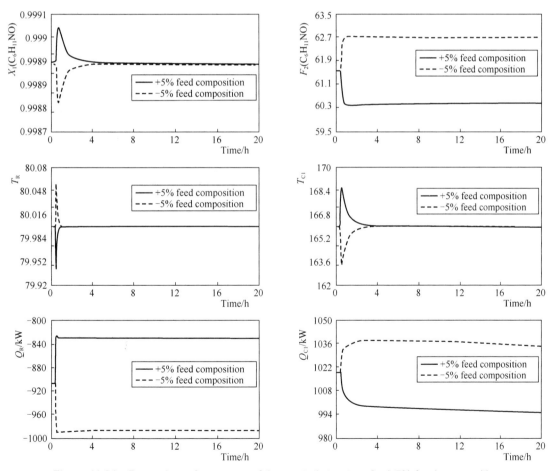

Figure 11.24 Dynamic performances of the control structure for ±5% feed composition disturbances of H_2O_2

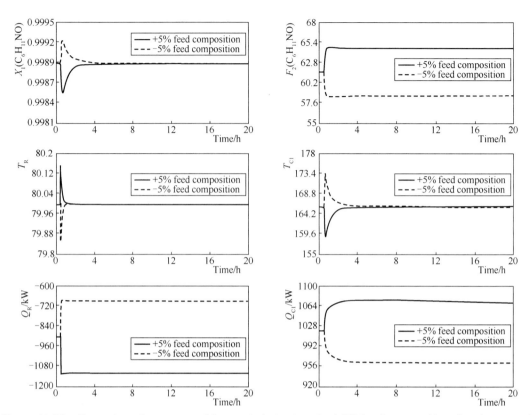

Figure 11.25　Dynamic performances of the control structure for ±5% feed composition disturbances of $C_6H_{10}O$

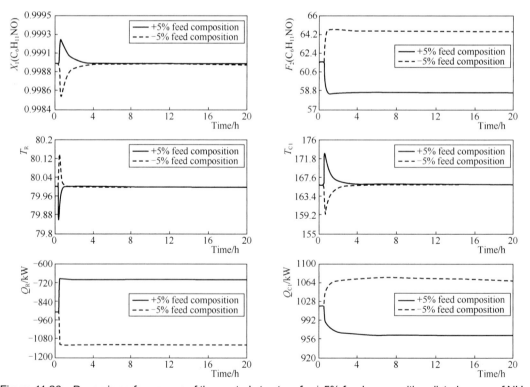

Figure 11.26　Dynamic performances of the control structure for ±5% feed composition disturbances of NH_3

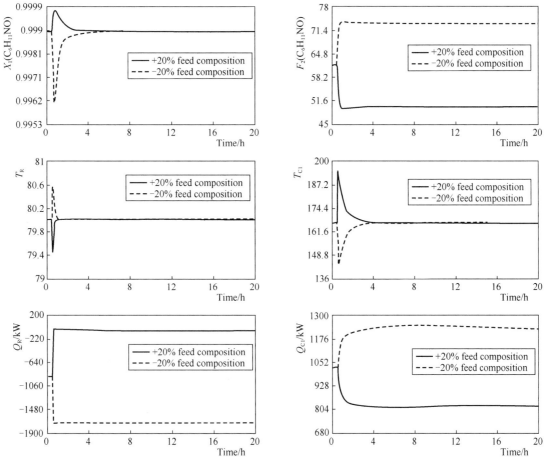

Figure 11.27 Dynamic performances of the control structure for ±20% feed composition disturbances of NH_3

When the disturbances are added, the conversion rate is nearly 1, as shown in Figure 11.28. Within a certain range, the conversion of $C_4H_{10}O$ is high. The component disturbances do not exceed this range. When flow disturbances occur, the composition of the feed does not change, so the conversion remains nearly 1. When a disturbance is added at 0.5 hours, there is a lag effect due to the original concentration distribution in the reactor. After a short period of time, the conversion returns to near 1.

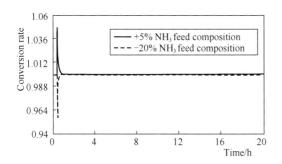
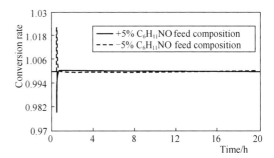

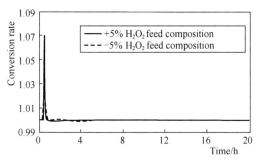 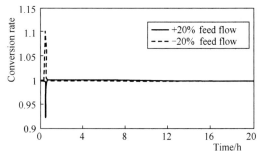

Figure 11.28 The effects of the feed flow rate and feed composition disturbances on the conversion

Exercises

1. The intrinsic kinetics of the amination of ethanol to acetonitrile over Co/γ-Al$_2$O$_3$ catalyst is studied. The reaction section was designed by RPlug model. The main reaction in the process is eq. (11-8), and three side reactions are eq. (11-9)~eq. (11-11). The kinetic parameters are shown in eq. (11-12)~eq. (11-15). The yield of the product is over 99.9%. Complete steady-state simulation and dynamic simulation.

$$C_2H_5ON + NH_3 \longrightarrow CH_3CN + H_2O + 2H_2 \tag{11-8}$$

$$CH_3CH_2OH \longrightarrow CH_3CHO + H_2 \tag{11-9}$$

$$CH_3CH_2OH + NH_3 \longrightarrow CH_3CH_2NH_2 + H_2O \tag{11-10}$$

$$2CH_3CH_2OH + NH_3 \longrightarrow CH_3CH_2CH_2\equiv N + 2H_2 + 2H_2O \tag{11-11}$$

$$r_{\text{acetonitrile}} = 1.06 \times 10^3 e^{-(67.38/RT)} C_{\text{ethanol}}^{-1.13} C_{\text{ammonia}}^{1.79} \tag{11-12}$$

$$r_{\text{acetaldehyde}} = 3.42 \times 10^8 e^{-(126.21/RT)} C_{\text{ethanol}}^{0.22} C_{\text{ammonia}}^{1.79} \tag{11-13}$$

$$r_{\text{ethylamine}} = 3.49 e^{-(6.92/RT)} C_{\text{ethanol}}^{1.13} C_{\text{ammonia}}^{-1.87} \tag{11-14}$$

$$r_{\text{butyronitrile}} = 1.54 e^{-(2.32/RT)} C_{\text{ethanol}}^{0.5} C_{\text{ammonia}}^{-0.17} \tag{11-15}$$

2. The kinetics of transesterification of methyl acetate with butanol is studied. The main reaction in the process is eq. (11-16). The kinetic equation is eq. (11-17). The kinetic parameters are shown in eq. (11-18) and eq. (11-19). The yield of the product is over 99.9%. Complete steady-state simulation and dynamic simulation.

$$MeAc + BuOH \rightleftharpoons MeOH + BuAc \tag{11-16}$$

$$R = K_F C_{\text{MeAc}} C_{\text{BuOH}} - K_R C_{\text{MeOH}} C_{\text{BuAc}} \tag{11-17}$$

$$K_F = 7 \times 10^6 e^{-71960/RT} \tag{11-18}$$

$$K_R = 9.467 \times 10^6 e^{-72670/RT} \tag{11-19}$$

References

[1] Zhu Z, Li G, Yang J, et al. Improving the energy efficiency and production performance of the cyclohexanone ammoximation process via thermodynamics, kinetics, dynamics, and economic analyses[J]. Energy Conversion and Management, 2019, 192: 100-113.

[2] Shen C, Wang Y J, Dong C, et al. In situ growth of TS-1 on porous glass beads for ammoximation of cyclohexanone[J]. Chemical Engineering Journal, 2014, 235: 75-82.

[3] Dong C, Wang K, Zhang J S, et al. Reaction kinetics of cyclohexanone ammoximation over TS-1 catalyst in a microreactor[J]. Chemical Engineering Science, 2015, 126: 633-640.

[4] Luyben W L. Distillation design and control using Aspen simulation[M]. John Wiley & Sons, 2013.

[5] Luyben W L. Evaluation of criteria for selecting temperature control trays in distillation columns[J]. Journal of Process Control, 2006, 16(2): 115-134.

[6] Tyreus B D, Luyben W L. Tuning PI controllers for integrator/dead time processes[J]. Industrial & Engineering Chemistry Research, 1992, 31(11): 2625-2628.

[7] Zhang D. Intrinsic kinetics for amination of ethanol to acetonitrile over Co-Ni/γ-Al_2O_3 catalyst[D]. Tian jin: Hebei University of Technology, 2011.